SEISMIC ANISOTROPY IN THE EARTH

MODERN APPROACHES IN GEOPHYSICS

formerly *Seismology and Exploration Geophysics*

VOLUME 10

The titles published in this series are listed at the end of this volume.

SEISMIC ANISOTROPY IN THE EARTH

by

V. BABUSKA
Institute of Geophysics, Prague

and

M. CARA
Institut de Physique du Globe, Strasbourg

KLUWER ACADEMIC PUBLISHERS
DORDRECHT / BOSTON / LONDON

Library of Congress Cataloging-in-Publication Data

Babuška, Vladislav.
Seismic anisotropy in the Earth / by V. Babuska, M. Cara.
p. cm.
Includes bibliographical references and index.
ISBN 0-7923-1321-6 (HB : acid free paper)
1. Seismic waves. 2. Anisotropy. 3. Earth--Interior. I. Cara, M. (Michel) II. Title.
QE538.5.B23 1991
551.2'2--dc20 91-20564

ISBN 0-7923-1321-6

Published by Kluwer Academic Publishers,
P.O. Box 17, 3300 AA Dordrecht, The Netherlands.

Kluwer Academic Publishers incorporates
the publishing programmes of
D. Reidel, Martinus Nijhoff, Dr W. Junk and MTP Press.

Sold and distributed in the U.S.A. and Canada
by Kluwer Academic Publishers,
101 Philip Drive, Norwell, MA 02061, U.S.A.

In all other countries, sold and distributed
by Kluwer Academic Publishers Group,
P.O. Box 322, 3300 AH Dordrecht, The Netherlands.

Printed on acid-free paper

Printed in the Netherlands

Contents

Preface

Structural geologists are well aware of the fact that isotropic rocks are quite exceptional in nature. Whichever origin, sedimentary, metamorphic or magmatic, rocks are shaped with a plane of mineral flattening, the foliation in geologists' jargon, and with a line of mineral elongation, the lineation. Just like a good quarryman, a trained structural geologist will detect a preferred orientation in an apparently isotropic granite. Preferred mineral orientation and thus structural anisotropy are the rule in nature. Considering the large variations in elastic coefficients of rock-forming minerals, it could be predicted that, in turn, seismic anisotropy should exist and be important, provided that domains with a similar structural signature are large enough to affect seismic waves.

This is why, in 1982 at a conference held in Frankfurt, which was one of the first meetings devoted to the subject of seismic anisotropy, I asked Don Anderson the question of why seismologists had not considered earlier in their models the obvious constraint of anisotropy. I still remember Don's answer: "Adolphe, we knew that our isotropic models were not very good but we had no other choice. It is simply that, so far, computers were not large enough to integrate the anisotropy parameter".

Changing isotropic glasses for anisotropic ones permits us to obtain better and more realistic seismic models of the Earth's interior, but, maybe more importantly, it has, for a seismologist, the far reaching consequence of stepping into the field of geodynamics. Since the measured seismic anisotropy has a structural origin, it brings information on large-scale structural features and can contribute to structural and geodynamic reconstructions. This is particularly clear in the upper mantle where seismic anisotropy reflects plastic flow either active or frozen, depending on whether seismic waves penetrate the asthenosphere or remain in the lithosphere. Thus seismic anisotropy maps, such as those of the Pacific ocean, which are presented in this volume can be converted into present and past mantle flow patterns below this ocean.

Considering the novelty of the subject and its potential developments, this monograph, written by two well-known specialists in seismic anisotropy, is welcome. It is altogether an introduction to the subject for a community extending beyond that of seismology, a remarkably documented review on seismic anisotropy and a thoughful discussion on how this complex subject can be interpreted. Coming at a time when data on seismic anisotropy is becoming increasingly abundant, Vladislav Babuska and Michel Cara's monograph extends the field of seismology from how the Earth is to how the Earth functions.

A. Nicolas

Montpellier, 29 January, 1991

Mathematical notation

General notation:

Scalars	a, α
Vectors	$\mathbf{a}$
Matrices	$\mathbf{A}$
Matrix elements	A_{ij}
Inverse of a matrix A	$\mathbf{A}^{-1}$
Transposition of a matrix A	$\mathbf{A}^{t}$

Specificic notation:

t	Time
∂_t	Time derivative
ω	Circular frequency
$\hat{u}(\omega)$	Fourier transform of u(t)
$\mathbf{r}$	Position vector
i, j, k, l	Indices of space coordinates (implicitly running from 1 to 3)
∂_i	Space derivative with respect to coordinate i
C_{ij}	Stiffness matrix elements
S_{ij}	Compliance matrix elements
c_{ijkl}	Stiffness tensor components

Chapter 1 - Introduction

Seismic waves are the most powerful tool geophysicists have at their disposal to investigate the *in situ* physical properties of the Earth's deep interior. The theory of elasticity is the basic theoretical framework upon which seismological methods are developed. The theory was already well advanced in the mid-nineteenth century, before it was proven that the Earth was a *quasi* perfectly elastic body for short time constants. The mechanical behavior of a linear elastic material depends primarily on Hooke's law which relates the stress to the strain. After the setting of the first general equations of elasticity by Navier, Cauchy, and Poisson in the 1820s, a controversy arose on the number of independent parameters needed to relate the stress to the strain. It was finally concluded, after the work of Green and Stokes in the 1830s and 1840s, that 21 independent elastic coefficients were required to describe the elastic properties of a solid in the most general case. This number reduces to 2 for isotropic solids and to 1 for fluids. The basic theoretical framework of the theory of elastic-wave propagations had then been set up a long time before geophysicists started exploring the Earth's interior with seismic waves.

Until the 1970s, the development of seismology and of its applied counterpart, seismic exploration, was mainly based upon two strong physical assumptions, that of perfect elasticity and that of elastic isotropy. Both assumptions are too restrictive to model the propagation of seismic waves in the real Earth.

Anelasticity was generally investigated independently of the study of seismic velocities. There are two main reasons for this. First, the attenuation of seismic waves is weak in the frequency band where they are observed, so that linear perturbation methods can be used by adding a small dissipative part in the elasto-dynamic equations. Second, as no physical dispersion is clearly apparent in the seismic observations, it was often supposed in the past that it was not necessary to correct the seismic velocities observed at different frequencies for the physical dispersion due to the dissipation of seismic energy. In fact seismology covers a broad frequency range, from 1 Hz to 1 mHz in the case of mantle seismic waves and physical dispersion is now observable at these frequencies. The largest departure from perfect elasticity is found in the asthenosphere at depths between 100 and 400 km, so that a correction for physical dispersion may be important for mantle waves.

The other strong *a priori* assumption, still commonly made in most seismic investigations, is that of elastic isotropy. In isotropic media seismic-wave velocities are independent of their direction of propagation and their polarization only depends on the type of wave which is considered, either of the compressional or shear types for body waves and of the Love or Rayleigh type for surface waves. On the other hand, in anisotropic media, seismic velocities depend locally on the direction of propagation and their polarization depends not only on the type of wave but also on the local symmetry of the elastic properties. Anisotropy of the Earth's materials was and still is often ignored in seismological investigations, although the reasons for neglecting anisotropy are much less clear than for anelasticity.

Anisotropy of physical properties is inherent to minerals. Laboratory studies of rock samples show that many mineral assemblages exhibit anisotropic elastic properties. Olivine is a major constituent of the Earth's upper mantle. It contributes probably to more than 60% of the upper mantle materials. Olivine crystals are strongly anisotropic, with up to 25% of compressional-wave velocity anisotropy. The possibility of shape-induced anisotropy at large wavelengths for layered media or distribution of cracks with a preferred orientation is another example where anisotropic seismic behavior may show up.

Of course, one might think that on a scale of seismic wavelengths, random orientations of microscopic constituents of rocks make the Earth's materials globally isotropic. This might be the case if large-scale physical fields such as the gravity field, the tectonic stresses or the shearing deformation due to a convective upper mantle did not induce any preferred orientation of the microscopic features of mineral assemblages. This might be also the case if no preferred orientation mechanism acted in the shearing zone of intense deformation near tectonic faults within the lithosphere. Such possibilities cannot be *a priori* ruled out but it is very likely that there are regions of the Earth's interior where such an assumption of overall isotropy is nothing else but an oversimplified statement. It seems much more reasonable to presume that seismic anisotropy should be *a priori* more the rule than the exception in every situation where dynamical processes are active, or were active in the past, and where preferred orientation of anisotropic material can occur over large volumes.

Since anisotropy may involve considerable complications in seismological investigations, it has quite often been ignored so far. Inverse methods applied to seismic data, e.g. to constructing tomographic images of the mantle, already lead to undetermined systems of equations. Introducing the anisotropy largely increases the number of parameters to be resolved. Even if we pass from an isotropic solid to a medium presenting an overall hexagonal symmetry with a vertical axis of symmetry, indeed one of the simplest anisotropic models, the number of elastic parameters increases from two - the two Lamé's parameters - to five. Considering the difficulties in constraining three-dimensional isotropic images of the Earth when we use the existing data, it is understandable to see many seismologists rather reluctant to accept the idea of increasing the number of unknowns in their inverse problems .

Direct evidence of seismic anisotropy is indeed difficult to observe. Observations have been accumulated, however, during the past 30 years from different types of seismic waves sampling the Earth's interior from the crust to the inner core. The elastic anisotropy as observed by seismic waves is not the same as the intrinsic anisotropy observed in a pure mineral during a laboratory experiment. It results from the preferred orientation of features of different scales. We use the term *seismic anisotropy* in this monograph for anisotropic properties which show up on the scale of a seismic wavelength, as opposed to microscopic anisotropy which corresponds to the elastic properties of individual crystals or rock samples. Many observations show that this large-scale seismic anisotropy is relatively small, often only a few per cent, and it is thus difficult to detect. However, even if it is small, we will see that hidden effects of anisotropy may be important when fine details of the Earth's structure are investigated. A trade-off between seismic anisotropy and other characteristics of Earth materials, such as the lateral or vertical variations of seismic velocity or the density distribution, may occur. When looking at the Earth with "isotropic glasses", we may encounter serious problems if some anisotropy is present somewhere along the seismic-wave path. Even in the case of weak anisotropy, very drastic changes in the basic properties of the seismic wavefield may appear: the most obvious example is the behavior of shear-wave polarization in anisotropic media.

Taking into account elastic anisotropy in seismological investigations appears thus necessary for many reasons. In addition to the obvious problems of trade-off which can occur between heterogeneities and anisotropy, studying seismic anisotropy is probably the only method today which can bring information on orientations of crystals and different types of fine structures under the action of large-scale geodynamical processes within the Earth's deep interior. It is not exaggerated to say that seismic anisotropy can be considered the most promising method of studying the deep tectonics and dynamics of the Earth.

This monograph is aimed at introducing the subject of seismic anisotropy to researchers and students in the fields of seismology, tectonophysics and geology. In this respect, one of its main goals is to facilitate the access to a quite complex literature dealing with the subject. The second objective of the monograph is to provide the reader with an up-to-date review of the seismic evidence and controversies related to the reality of the seismic anisotropy within the Earth. Finally, we have tried to develop in these pages a philosophy to deal with future seismic experiments and interpretations of seismological observations in terms of anisotropic properties of the Earth's materials.

Considering the very broad topic covered by this monograph, we could not deal with all aspects of seismic anisotropy. In particular, we have focused our attention more on deep geodynamic processes than on the upper crystalline crust and sedimentary rocks, despite the potentially very promising applications of seismic anisotropy to estimating some of the properties of the seismogenic zones and oil reservoirs. As regards theoretical aspects of wave propagation in anisotropic media, our intention was mainly to provide the reader with an introduction to the subject. For this reason, we have limited the theoretical developments to several illustrative examples and to a discussion where the main results are given with reference to the specialized literature.

Chapter 2 is a review of the properties of seismic wave propagation in anisotropic media where the fundamental concepts of elastic anisotropy and the principal properties of seismic waves in anisotropic media are given. Chapter 3 deals with laboratory studies of rock-forming minerals and polycrystalline rocks. More specific questions of interpreting seismic observations in terms of anisotropy are presented in chapter 4. Chapter 5 reviews the present evidence of seismic anisotropy in the lithosphere, including the crust, and chapter 6 deals with the seismic observations and models concerning the boundary between the lithosphere and the asthenosphere and the deeper parts of the Earth.

This monograph is based on numerous discussions both authors have had with many scientists during the last ten years. We are particularly indebted to V. Ansel, S. Crampin, Y. Gueguen, H. Kern, K. Klima, R.C. Liebermann, D. Mainprice, V. Maupin, J.P. Montagner, A. Nicolas, J. Plomerova, V. Pribylova, I. Psencik and V. Vavrycuk for their criticism and suggestions while we were preparing the manuscript.

Prague, March 29, 1991

Chapter 2 - Seismic-wave propagation in anisotropic media

Seismic-wave propagation in anisotropic media is generally considered to be a difficult subject. In an anisotropic medium, the seismic velocities vary with the direction of propagation. The polarization of a given seismic wave depends not only on the type of the wave and its direction of propagation, but also on the elastic properties of the material. Differences between phase and group velocities of the non-dispersed body waves is one additional complication. All these properties have been thoroughly studied for crystals and rather extensive mathematics are necessary, indeed, to deal with these subjects.

In fact, the complexity of the subject becomes really serious when we deal with strongly anisotropic materials presenting a low degree of symmetry. We will see, for example, that a fourth-order tensor with 21 independent elastic coefficients is necessary to describe a material with triclinic symmetry, the lowest degree of symmetry. On the scales of interest for seismic waves in the Earth, i.e. for wavelengths ranging from several tenths of a meter in oil seismic exploration techniques to hundreds of kilometers in long-period seismology, it is very likely that the departure from isotropy is rather small, so that simplified methods based on the first-order perturbation theory can be generally used. Even though some drastic changes of the seismic wavefield may occur, there are several situations where the behavior of the seismic waves remains close enough to what is observed in isotropic structures to make fairly simple methods of interpretation workable.

As we will see, the theory of elastic anisotropy is rather straightforward to develop in homogeneous media. In heterogeneous media, a question of fundamental importance is that the seismic wave properties depend on the wavelength. Apparent anisotropy may for instance appear on the macroscopic scale even if the materials are intrinsically isotropic but with preferentially oriented heterogeneities. In order to introduce this fundamental concept, let us first start with a simple example.

2.1 A simple anisotropic medium: the stack of isotropic layers

It is well known that the overall elastic properties of a heterogeneous medium, such as those affecting long-wavelength seismic waves, differ from those observed on a small scale, say on the scale of a crystal in polycrystalline aggregates. The key question for the existence of an overall anisotropy is the preferred orientation of the microscopic features in the material. A stack of parallel isotropic layers presenting alternate high and low rigidities is a simple example exhibiting such a behavior.

For the sake of illustration, let us consider two experiments where a stack of homogeneous isotropic layers is elastically deformed under static shearing stresses as depicted in Fig. 2-1a and 2-1b. Let us call $\mathbf{u}(x,y,z)$ the displacement vector field which is associated with the elastic deformation. In both cases, the following relationships between the shearing stress S_n in layer n, the local shearing deformations $\partial u_x / \partial y$ and $\partial u_y / \partial z$ and the local rigidities μ_n hold:

$$\partial u_x / \partial y = S_n / \mu_n \quad n=1,2;$$

$$\partial u_y / \partial z = S_n / \mu_n \quad n=1,2.$$

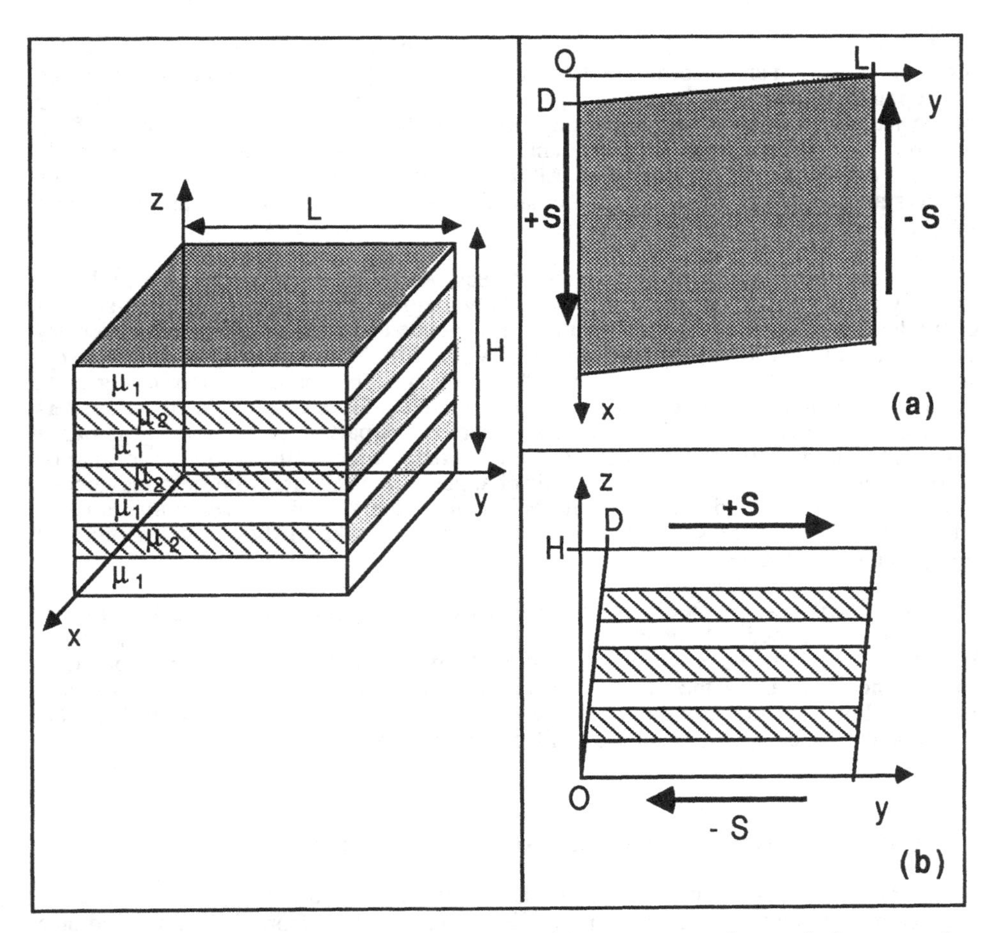

Figure 2-1 Elastic deformations of a stack of homogeneous isotropic layers under two types of shearing stresses. The overall material is stiffer when deformed as in (a) than in (b). To memorize the situation, a trick may be used. Imagine that you hold in your hand a stack of playing cards: deformation is easy in situation (b) where the cards slip parallel to each other while resistance to deformation is strong in situation (a) due to the rigidity of the cards.

The boundary conditions are not the same in Fig. 2-1a and 2-1b. In (a), the geometry of the problem leads to a uniform displacement field at the two plane boundaries $y = 0$ and $y = L$ where the shearing forces S and -S are applied. The high rigidity layers suffer from a high stress S_1 while the low-rigidity layers present a low stress S_2.

The average shearing stress of the medium is:

$$<S> = (S_1 e_1 + S_2 e_2) / H,$$

where H is the total thickness of the stack of layers and e_1 and e_2 are the cumulative thicknesses of the layers of type 1 and 2, respectively. Replacing S_1 and S_2 by their respective values we get:

$$<S> = (\mu_1 e_1 / H + \mu_2 e_2 / H)\, \partial u_x / \partial y\,, \quad \text{or} \quad <S> = <\mu> D / L\,,$$

where D is the displacement shown in Fig. 2-1a and L is the length of the layers taken along the y axis. The apparent rigidity $<\mu>$ of the medium thus appears as a linear average of the local rigidities:

$$<\mu> = \mu_1 e_1 / H + \mu_2 e_2 / H\,.$$

This type of average corresponds to Voigt's average (Voigt, 1928).

In Fig. 2-1b, on the other hand, the shearing stress is applied uniformly to the upper and lower boundaries of the stack of layers, so that the stress is uniform within all the material while the local strain $\partial u_y / \partial z$ varies from point to point:

$$S = \mu_1\, \partial u_{1y} / \partial z = \mu_2\, \partial u_{2y} / \partial z\,.$$

The overall strain D / H can be computed by dividing the total relative displacement D by the total thickness H of the stack of layers:

$$D / H = (e_1\, \partial u_{1y} / \partial z + e_2\, \partial u_{2y} / \partial z) / H = (e_1 / \mu_1 + e_2 / \mu_2)\, S / H.$$

If we define a new apparent rigidity $<\mu'>$ as

$$1/ <\mu'> = (e_1 / \mu_1 + e_2 / \mu_2) / H\,,$$

we obtain the following relationship between the stress S and the overall strain D / H:

$$S = <\mu'> D/H.$$

Inverses of rigidities add linearly in this case. This means that the compliances add linearly. This is Reuss' average (Reuss, 1929) which leads to smaller apparent rigidities than Voigt's average.

A stack of homogeneous isotropic layers thus behaves as an anisotropic material if overall apparent elastic properties are considered. If, for example, a seismic S wave with a wavelength much larger than the layer thicknesses propagates in this stack of layers, its velocity will depend on its polarization and direction of propagation since two different rigidities $<\mu>$ and $<\mu'>$ are involved, depending on the type of shearing deformation. A more complete demonstration of the properties of a stack of thin laminated layers of variable thicknesses was presented by Backus (1962).

Anisotropic properties of stacked isotropic layers have been known for a long time (Riznichenko, 1949; Thomson, 1950; Postma, 1955). They have been observed in seismic refraction works in the 1950's (Uhrig and Van Melle, 1955; Layat et al., 1961). Laboratory measurements on rock samples have also shown that oil shale presents such a type of anisotropy (Kaarsberg, 1968). In seismology, Toksoz and Anderson (1963) applied such a simple anisotropic model in order to obtain the properties of a "transversely isotropic" spherical mantle: for this purpose they designed an acoustic experiment using analogous seismic models with parallel grooves engraved in an aluminium disc.

More generally, several averaging processes can be used to define the overall elastic properties of a material on the macroscopic scale. Usual averaging processes are generally referred to as Voigt's (1928) and Reuss's (1929) averages. Voigt's average is computed by averaging the local stiffnesses of the material, likes the rigdities μ_n in the above example, while Reuss' average concerns the compliances (e.g. $1/\mu_n$ in the above example). A third type of average, Hill's average, is also frequently applied. It corresponds to the average of Voigt's and Reuss' averages (Hill, 1952).

2.2 Elastic tensors in general anisotropic media

The rigidities $<\mu>$ and $<\mu'>$ introduced in the previous section are two components of a more general tensor relating the elastic strain to the stress: the stiffness tensor. Let us now examine the general form of this tensor for an arbitrary anisotropic medium. The stress and strain fields are second-order symmetric tensor fields. The general linear relationship between these two tensors is known as Hooke's law. It involves a fourth-order elastic tensor, either the stiffness tensor relating the strain to the stress or its inverse, the compliance tensor, relating the stress to the strain. To analyse the symmetry properties of these elastic tensors, it is necessary to base Hooke's law on thermodynamics. In a fluid, the pressure is related to the partial derivative of the internal energy with respect to variations of the volume. Likewise, in elastic solids, the stress tensor components σ_{ij} are related to the partial derivative of the density of internal elastic energy U with respect to the strain tensor component ε_{ij} :

$$\sigma_{ij} = \partial U / \partial \varepsilon_{ij}$$

Developing the elastic energy U of a continuous medium into powers of the strain tensor components, and stating that the stress is zero in the undeformed state, we find that the density of energy U is a quadratic function of the strain tensor components for small deformations (e.g. Landau and Lifshitz, 1964). Making use of the Einstein summation rule for repeated lower indices, one can thus write in any Cartesian coordinate system:

$$U = c_{ijkl}\, \varepsilon_{ij}\, \varepsilon_{kl} / 2,$$

where c_{ijkl} is defined as the components of the fourth-order stiffness tensor depending on the local elastic properties of the material.

Computing the partial derivative $\partial U / \partial \varepsilon_{ij}$, one gets the general form of Hooke's law:

$$\sigma_{ij} = c_{ijkl}\, \varepsilon_{kl}.$$

The symmetries of both the stress and deformation tensors allow the permutation of indices i and j on the one hand, and of k and l, on the other. The existence of a density energy function U which is quadratic in strain furthermore allows the permutation of the two pairs of indices (i,j) and (k,l). We can thus write:

$$c_{ijkl} = c_{jikl} = c_{ijlk} = c_{klij}\ .$$

The above symmetry conditions allow us to reduce the number of independent elastic coefficients from 81 to 21 in an arbitrary anisotropic medium.

To show this, let us first notice that the symmetry condition $c_{ijkl} = c_{jikl}$ leads to consider six independent pairs (i,j). Likewise, since $c_{ijkl} = c_{ijlk}$ there are six independent pairs (k,l). Taking advantage of this symmetry, anisotropic media are often described by using a 6×6 matrix C_{ij} instead of the full c_{ijkl} tensor components. Let us stress that this matrix is defined on purely conventional grounds and has no physical basis: the description of the elastic properties of a solid needs a fourth-order tensor, it cannot be restricted to a second-order tensor even if for some practical reasons it is convenient to order the 36 remaining components in a two-dimensional table C_{ij}.

The usual definition of matrix C_{ij} is such that the indices i and j of C_{ij} vary from 1 to 6 if the pairs of indices (i,j) or (k,l) of the elastic tensor c_{ijlk} take values (1,1), (2,2), (3,3), (2,3), (1,3) and (1,2), respectively (Voigt, 1928).

More formally, $C_{ij} = c_{klmn}$ with $i=k=l$ if $k=l$ and $i=9-k-l$ if $k \neq l$, and $j=m=n$ if $m=n$ and $j=9-m-n$ if $m \neq n$,

or explicitly:

$$(C_{ij}) = \begin{pmatrix} c_{1111} & c_{1122} & c_{1133} & c_{1123} & c_{1113} & c_{1112} \\ c_{2211} & c_{2222} & c_{2233} & c_{2223} & c_{2213} & c_{2212} \\ c_{3311} & c_{3322} & c_{3333} & c_{3323} & c_{3313} & c_{3312} \\ c_{2311} & c_{2322} & c_{2333} & c_{2323} & c_{2313} & c_{2312} \\ c_{1311} & c_{1322} & c_{1333} & c_{1323} & c_{1313} & c_{1312} \\ c_{1211} & c_{1222} & c_{1233} & c_{1223} & c_{1213} & c_{1212} \end{pmatrix}$$

Finally, the symmetry condition $c_{ijlk} = c_{lkij}$ allows us to eliminate (36-6)/2 =15 out of the 36 remaining coefficients of the above matrix since the matrix is symmetric. In the most general anisotropic solid (triclinic symmetry), the number of independent elastic coefficients can thus be reduced to 21. The number of independent coefficients can be reduced still more in a material presenting higher symmetry than the triclinic one. In fluids which make the end of this family presenting fewer and fewer independent coefficients, the fourth-order stiffness tensor c_{ijlk} is reduced to one elastic coefficient, the bulk modulus k:

$$c_{ijlk} = k \, \delta_{ij} \, \delta_{kl} \, ,$$

δ_{ij} being the Kronecker symbol (1 if i=j, 0 if i≠j). In isotropic solids, two independent elastic coefficients remain: Lamé's coefficients usually denoted λ and μ. The stiffness tensor components of an isotropic solid are:

$$c_{ijkl} = \lambda \, \delta_{ij} \, \delta_{kl} + \mu \, (\delta_{ik} \, \delta_{jl} + \delta_{il} \, \delta_{jk}).$$

For the sake of illustration, we give in Table 2-1 the number of independent elastic coefficients which are necessary to describe the elastic tensor in different symmetry systems.

Table 2-1 Number of independent elastic coefficients for selected symmetry systems and typical minerals or Earth's materials.

Type of symmetry	Number of independent elastic coefficients	Typical mineral
triclinic	21	plagioclase
monoclinic	13	hornblende
orthorhombic	9	olivine
tetragonal	6	stishovite
trigonal I	7	ilmenite
trigonal II	6	quartz
hexagonal	5	ice
cubic	3	garnet
isotropic solid	2	volcanic glass

Fig. 2-2 gives the spatial orientations of typical symmetry planes in the more common symmetry systems. This figure shows, for example, that there is one plane of symmetry in the monoclinic case: monoclinic crystals are symmetric with respect to the plane (010) which is parallel to the two crystallographic axes [100] and [001] (or a- and c-axes).

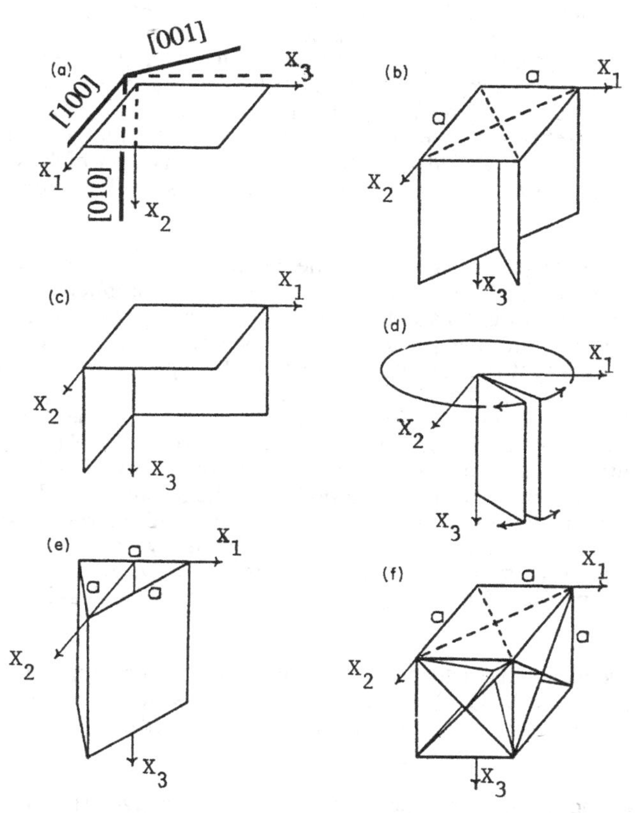

Figure 2-2 Example of symmetry planes in the different symmetry systems: (a) monoclinic, (b) tetragonal, (c) orthorhombic, (d) hexagonal, (e) trigonal and (f) cubic. Crystallographic axes are indicated in the monoclinic case (adapted from Crampin, 1984c; with permission of the Royal Astronomical Society, copyright © 1984).

When computing the Cartesian components of the stiffness tensor, the mutually perpendicular axes (x_1,x_2,x_3) are generally oriented in such a way that the crystallographic axes coincide, if possible, with the Cartesian axes. This is, of course, not possible when the three crystallographic axes are not mutually perpendicular. For example, in the case of monoclinic crystals, the natural way of orienting the Cartesian frame is to set x_1 parallel to [100], x_2 parallel to [010] and x_3 perpendicular to the plane (001) containing the axes [100] and [010] (see Fig. 2-2). Additional information on this question can be found in Nye (1979) (see also the Table captions 3-2 and 3-3 in Chapter 3).

Note that the number of independent parameters must not be confused with the number of non-zero coefficients in the remaining 21 elastic components of the elastic stiffness tensor. In the case of the triclinic symmetry for example, the number of non-zero independent elastic parameters can be reduced from 21 to 18 by choosing an appropriate coordinate system (3 angles are necessary to orient the reference frame). In the monoclinic case, the symmetry with respect to one plane leads to reducing the number of elastic coefficients from 21 to 13, but rotating the frame within this symmetry plane may allow us to set to zero one of the coefficients and 12 non-zero parameters are then left (Landau and Lifschitzt, 1964). Orthorhombic symmetry which corresponds to symmetry with respect to three orthogonal planes leads to 9 independent elastic coefficients.

Finally let us describe in greater detail the hexagonal symmetry case which is of considerable practical interest in seismology due to its relative simplicity and its ability to approximate many actual situations in the Earth. First of all, let us note that hexagonal symmetry and cylindrical symmetry around one single axis are equivalent for the elastic properties. This means that the elastic properties of hexagonal crystals are invariant when rotated around their 6-fold axis. Similarly any elastic material exhibiting invariant properties when rotated around one fixed direction in space can be described by the elastic tensor of a hexagonal system. Since the elastic properties of bodies exhibiting hexagonal symmetry are invariant in the plane perpendicular to the axis of symmetry, they are often called transversely isotropic bodies in the seismological literature after Love (1927). This type of symmetry is also called azimuthal isotropy, axi-symmetry, cylindrical symmetry or even radial symmetry if the symmetry axis is vertical (i.e. radial away from the center of the spherical Earth). Detail review of seismic wave properties in weakly anisotropic media with a symmetry around the vertical axis was given by Thomsen (1986).

There are 5 independent elastic coefficients in the hexagonal symmetry case. Let us mention the coefficients A, C, F, L, N (Love, 1927), which are often used in the seismological literature. Love's notations are related to the stiffness tensor components

Table 2-2 Five independent elastic coefficients of a hexagonal symmetry system (x_3 is the symmetry axis of the material).

Love (1927)	Matrix notations	Stiffness tensor components
A	$C_{11} = C_{22}$	$c_{1111} = c_{2222}$
C	C_{33}	c_{3333}
F	$C_{31} = C_{32}$	$c_{3311} = c_{3322}$
L	$C_{44} = C_{55}$	$c_{2323} = c_{1313}$
N	C_{66}	c_{1212}
Dependent coefficients:	$C_{12} = C_{21} = C_{11} - 2\,C_{66}$	$c_{1122} = c_{2211} = c_{1111} - 2\,c_{1212}$

in Table 2-2 for an axis of symmetry parallel to axis x_3. In order to describe the elastic properties of such a material within the Earth, it is necessary to orient the symmetry axis in space, with respect to the geographical coordinate frame for instance. Thus we need five coefficients A, C, F, L, N, plus two angles to orient the symmetry axis when describing the *in situ* elastic properties.

The layered structure presented in the previous section is an example of an overall anisotropic material exhibiting cylindrical symmetry around the Cartesian axis x_3. Cracked isotropic media are another type of media which may exhibit anisotropic properties with cylindrical symmetry (Nur, 1971; Anderson et al., 1974). Much attention has been recently given to such media because they are of great interest in the seismic exploration of oil fields and the study of earthquake source regions (Crampin, 1987). The properties of cracked media are also of importance in the study of the tectonic stress field in the deep brittle crustal layers, since a stress field can open preexisting cracks parallel to the direction of maximum compressive stress while cracks perpendicular to this direction may remain closed. A review of the seismic properties of cracked media has been given by Crampin (1984a). Hudson (1980, 1981) developed the theory of seismic wave propagation in cracked media with different boundary conditions at the crack boundaries.

In the simple case of fluid filled cracks with an infinite aspect ratio, i.e. flat ellipsoids, Hudson's expressions are similar to those derived by Garbin and Knopoff (1975) for weak anisotropy. If the normals to the plane of the cracks are oriented parallel to axis x_1, the stiffness matrix C_{ij} for flat fluid filled cracks may be written as:

$$(C_{ij}) = \begin{pmatrix} \lambda+2\mu & \lambda & \lambda & 0 & 0 & 0 \\ \lambda & \lambda+2\mu & \lambda & 0 & 0 & 0 \\ \lambda & \lambda & \lambda+2\mu & 0 & 0 & 0 \\ 0 & 0 & 0 & \mu & 0 & 0 \\ 0 & 0 & 0 & 0 & \mu(1-\varepsilon) & 0 \\ 0 & 0 & 0 & 0 & 0 & \mu(1-\varepsilon) \end{pmatrix}$$

where λ and μ are the Lamé's coefficients of the isotropic medium, $\varepsilon = N a^3 / V$ is the crack density, N being the number of cracks in the volume V and "a" being the radius of the cracks.

The above stiffness matrix corresponds to a particularly simple case of hexagonal symmetry with the symmetry axis perpendicular to the plane of preferred orientation of the cracks. In case of a more general crack geometry, the degree of symmetry is lower. Hudson (1990) gives the stiffness matrices of the equivalent overall anisotropic media for arbitrary crack geometry.

In summary, using the matrix form C_{ij} we present in Table 2-3 the elastic coefficients for two cases of symmetry which play a major role on the large scale in geophysics: the orthorhombic and the hexagonal symmetries. The isotropic case is also shown for comparison purposes. Matrices C_{ij} in other symmetry systems are given, e.g., by Nye (1979) and Crampin (1989). Examples of matrices C_{ij} for the most common mineral of the crust and upper mantle are given in Chapter 3 (see Tables 3-2 and 3-2).

Table 2-3 Elastic coefficients of orthorhombic (upper left), hexagonal (upper right) and isotropic materials (*).

$$(C_{ij}) = \begin{pmatrix} a & b & c & \cdot & \cdot & \cdot \\ b & d & e & \cdot & \cdot & \cdot \\ c & e & f & \cdot & \cdot & \cdot \\ \cdot & \cdot & \cdot & g & \cdot & \cdot \\ \cdot & \cdot & \cdot & \cdot & h & \cdot \\ \cdot & \cdot & \cdot & \cdot & \cdot & i \end{pmatrix} \qquad (C_{ij}) = \begin{pmatrix} A & A\text{-}2N & F & \cdot & \cdot & \cdot \\ A\text{-}2N & A & F & \cdot & \cdot & \cdot \\ F & F & C & \cdot & \cdot & \cdot \\ \cdot & \cdot & \cdot & L & \cdot & \cdot \\ \cdot & \cdot & \cdot & \cdot & L & \cdot \\ \cdot & \cdot & \cdot & \cdot & \cdot & N \end{pmatrix}$$

$$(C_{ij}) = \begin{pmatrix} \lambda+2\mu & \lambda & \lambda & \cdot & \cdot & \cdot \\ \lambda & \lambda+2\mu & \lambda & \cdot & \cdot & \cdot \\ \lambda & \lambda & \lambda+2\mu & \cdot & \cdot & \cdot \\ \cdot & \cdot & \cdot & \mu & \cdot & \cdot \\ \cdot & \cdot & \cdot & \cdot & \mu & \cdot \\ \cdot & \cdot & \cdot & \cdot & \cdot & \mu \end{pmatrix}$$

(*) Notations:

- The zero coefficients are indicated by dots.
- In the orthorhombic case, the planes of symmetry are perpendicular to the Cartesian axes x_1, x_2, and x_3. Letters a, b, c, d, f, g, h, and i represent, accordingly, the following elastic coefficients:

$a = c_{1111}$, $b = c_{2211} = c_{1122}$, $c = c_{3311} = c_{1133}$,
$d = c_{2222}$, $e = c_{3322} = c_{2233}$, $f = c_{3333}$,
$g = c_{2323}$, $h = c_{1313}$ $i = c_{1212}$

- In the hexagonal case, axis x_3 is the axis of symmetry and A, C, F, L, N are the Love 1911's elastic coefficients (see Table 2-2).
- λ and μ are Lamé's coefficients (isotropic case).

In the above examples the stiffness matrices C_{ij} are given in Cartesian coordinates well adapted to the symmetry of the material. The computation of the stiffness tensor components in other coordinate systems requires the rotation of the coordinate axes using the rotation matrices and the rules of tensorial component transformations (e.g. Montagner and Anderson, 1989a, for the use of Euler's angles in computing elastic constants of polycrystalline aggregates).

Before closing this section on elastic tensors, let us mention that there is another approach to describing the elastic properties of a material. Instead of developing the theory in terms of the stiffness tensor components c_{ijkl} which relate the strain to the stress, it is possible to use the compliances which relate the stress to the strain. The easiest way to derive the compliances from the stiffnesses is to write Hooke's law in matrix notation. If we write the six independent strain and stress tensor components in column vectors with $\varepsilon_1 = \varepsilon_{11}$, $\varepsilon_2 = \varepsilon_{22}$, $\varepsilon_3 = \varepsilon_{33}$, $\varepsilon_4 = 2\,\varepsilon_{23}$, $\varepsilon_5 = 2\,\varepsilon_{31}$, $\varepsilon_6 = 2\,\varepsilon_{12}$, and $\sigma_1 = \sigma_{11}$, $\sigma_2 = \sigma_{22}$, $\sigma_3 = \sigma_{33}$, $\sigma_4 = \sigma_{23}$, $\sigma_5 = \sigma_{31}$, $\sigma_6 = \sigma_{12}$, Hooke's law may be written as:

$$\sigma_n = C_{nm}\,\varepsilon_m \,.$$

Inverting the stiffness matrix C_{nm} allows us to compute the compliances S_{mn} since:

$$\varepsilon_m = S_{mn}\,\sigma_n \,.$$

2.3 Plane waves in homogeneous anisotropic media

Let us now consider the simplest type of seismic waves in anisotropic media: plane waves propagating in a homogeneous anisotropic medium. If $\mathbf{u}(\mathbf{r},t)$ denotes the motion of a particle located at coordinates $\mathbf{r}$ at time t, one can write for a plane wave propagating in the direction $\mathbf{n}$ at phase velocity c:

$$\mathbf{u}(\mathbf{r},t) = \mathbf{a}\, f\,(t - \mathbf{n} \,.\, \mathbf{r} / c\,),$$

where the function f() gives the time dependence of the particle motion at a fixed point in space, **a** is a vector giving the amplitude and the polarity of the wave and **n** is a unit vector perpendicular to the phase surface.

The conditions to be satisfied by the vector **a** and the phase velocity c can easily be obtained from the equation of elastodynamics applied to a homogeneous medium. In Cartesian coordinates, one can write:

$$\partial_j\, \sigma_{ij} = c_{ijkl}\, \partial_j\,(\varepsilon_{kl}) = \rho\, \partial_t^2\,(\,u_i\,),$$

where ∂_j denotes the space derivative relative to the coordinate x_j, ∂_t denotes the time derivative and ρ is the density. For small strain, we have

$$\varepsilon_{kl} = (\partial u_k / \partial x_l + \partial u_l / \partial x_k) / 2,$$

and taking into account the symmetry of both the strain tensor and the elastic tensor, one can also write:

$$c_{ijkl}\, \partial_j \partial_k(\,u_l\,) = \rho\, \partial_t^2\,(u_i\,).$$

Finally, using the relation of the displacement field $\mathbf{u}(\mathbf{r},t)$, one gets:

$$m_{il}\, a_l = c^2\, a_i\,,$$

where the Christoffel matrix elements m_{il} are

$$m_{il} = c_{ijkl}\, n_j\, n_k \,/\, \rho.$$

The square of the phase velocity c appears as an eigenvalue of the Christoffel matrix (m_{il}) and the vectors **a** are the associated eigenvectors. Note that due to the symmetry of the stiffness tensor c_{ijkl}, the Christoffel matrix is symmetric so that, except in the case of degenerate eigenvalues, its eigenvectors are mutually perpendicular.

2.3.1 Hexagonal symmetry case

Let us solve this eigenvalue-eigenvector problem in the case of hexagonal symmetry when the seismic waves propagate either along the symmetry axis of the material or perpendicular to it. As noted above, this kind of anisotropy is widely used in the seismological literature. This simple example allows us to introduce the notion of S-wave splitting which is one of the most important properties of the seismic wavefield in anisotropic materials.

As above, we choose the Cartesian axis x_3 as an axis of symmetry. Let us examine a plane wave which propagates along the axis x_1. If the cartesian unit vectors are denoted $(\mathbf{e}_1, \mathbf{e}_2, \mathbf{e}_3)$, we have $\mathbf{n} = \mathbf{e}_1$. Simple computations then yield the Christoffel matrix which appears in the diagonal form:

$$(m_{ij}) = \frac{1}{\rho}\begin{pmatrix} A & 0 & 0 \\ 0 & N & 0 \\ 0 & 0 & L \end{pmatrix}$$

where A, N and L are the Love's (1911) elastic coefficients. If $L \neq N$, the Christoffel matrix has three distinct eigenvalues λ_p which are equal to its diagonal terms:

$$\lambda_1 = A/\rho, \;\; \lambda_2 = N/\rho \;\text{ and }\; \lambda_3 = L/\rho.$$

Three plane waves can thus propagate along axis x_1. These waves exhibit different velocities and polarizations. The unit eigenvectors $\mathbf{a}_n$, n = 1, 2, 3, associated with the eigenvalues λ_1, λ_2, and λ_3 , give the directions of polarization. They are, respectively:

$$\mathbf{a}_1 = \begin{vmatrix} 1 \\ 0 \\ 0 \end{vmatrix} \qquad \mathbf{a}_2 = \begin{vmatrix} 0 \\ 1 \\ 0 \end{vmatrix} \qquad \mathbf{a}_3 = \begin{vmatrix} 0 \\ 0 \\ 1 \end{vmatrix}$$

The fastest wave is associated with the eigenvalue λ_1. Its velocity is

$$\alpha_1 = \sqrt{\lambda_1} = \sqrt{\frac{A}{\rho}}$$

and it is polarized parallel to the direction of propagation (P_1wave in Fig. 2-3).

If we consider a medium where $N > L$ (e.g. a plane layered structure with alternatively high and low rigidity layers perpendicular to axis x_3 as in section 2.1), the second wave observed on a record propagates with phase velocity

$$\beta_1 = \sqrt{\lambda_2} = \sqrt{\frac{N}{\rho}}$$

and its polarization vector lies in the plane (x_1,x_2), parallel to axis x_2. This wave is labelled S_1 in Fig. 2-3. The slowest wave is labelled S_2 in Fig. 2-3. It propagates with

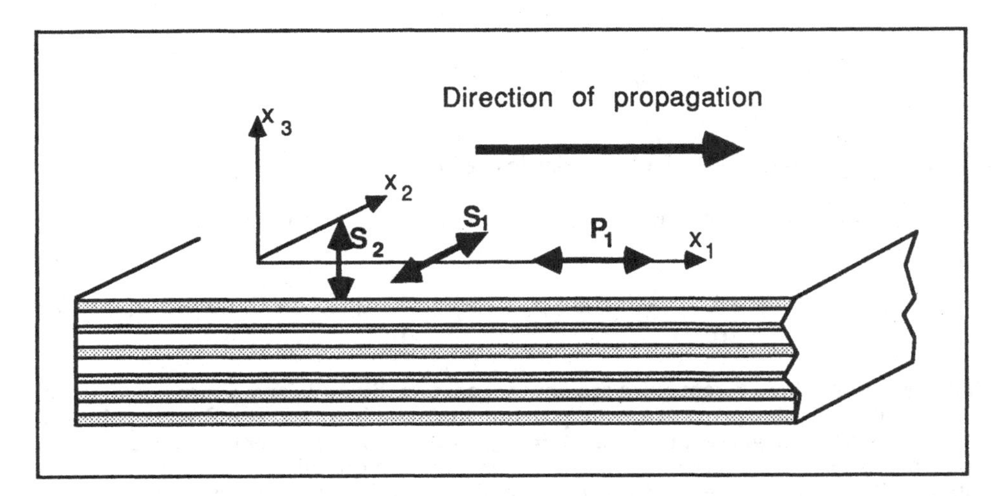

Figure 2-3 Three possible seismic plane-waves propagating in the plane perpendicular to the symmetry axis (x_3 here).

phase velocity

$$\beta_2 = \sqrt{\lambda_3} = \sqrt{\frac{L}{\rho}}$$

and is polarized parallel to the axis of symmetry x_3. Due to the symmetry around axis x_3, the above properties are invariant in the plane perpendicular to axis x_3, i.e. for any direction of propagation in this plane.

If the medium were isotropic, the eigenvalues λ_2 and λ_3 would be equal. This would correspond to a situation of degenerate eigenvalues where any vector in the plane perpendicular to the direction of propagation x_1 would become a possible direction of polarization for S waves. In this case, no preferred polarization direction would thus be observed for S waves. On the other hand, even if very small anisotropy is present, S waves propagating in directions perpendicular to the symmetry axis are split into two waves polarized perpendicularly, according to the direction of the eigenvectors of the Christoffel matrix. After an initial splitting, e.g. at $x_1=0$, the arrival-time difference δt between the waves S_1 and S_2 will grow with the distance x_1 as

$$\delta t = x_1 \, (1/\beta_2 - 1/\beta_1),$$

so that even if the anisotropy is small, the time difference δt can be noticeable after a long enough path.

Let us now consider a plane wave propagating along the axis x_3. Computation of the Christoffel matrix for $\mathbf{n} = \mathbf{e}_3$ yields:

$$(m_{ij}) = \frac{1}{\rho}\begin{pmatrix} L & 0 & 0 \\ 0 & L & 0 \\ 0 & 0 & C \end{pmatrix}$$

where, again, L and C are the Love's elastic coefficients.

The unit eigenvectors and associated eigenvalues can be chosen as

$$\lambda_1 = L/\rho \;,\; \mathbf{a}_1 = \begin{vmatrix} 1 \\ 0 \\ 0 \end{vmatrix} \;;\; \lambda_2 = L/\rho \;,\; \mathbf{a}_2 = \begin{vmatrix} 0 \\ 1 \\ 0 \end{vmatrix} \;;\; \lambda_3 = C/\rho \;,\; \mathbf{a}_3 = \begin{vmatrix} 0 \\ 0 \\ 1 \end{vmatrix} .$$

Two out of the three eigenvalues are now degenerate. Since λ_1 and λ_2 are equal, there are only two possible plane waves propagating in the x_3 direction: one P wave polarized parallel to the propagation direction (P_2 in Fig. 2-4) and one degenerate S wave with an arbitrary polarization perpendicular to the propagation direction (S_2 in Fig. 2-4). It is remarkable that S waves propagating in the direction of the symmetry axis x_3 correspond to the same eigenvalue L/ρ, and have thus the same velocity, β_2, as the S wave propagating at a right angle with a polarization parallel to the symmetry axis. Note also that the P wave propagating in the direction of the symmetry axis (P_2 in Fig. 2-4) corresponds to an eigenvalue C/ρ which is different from that of the P wave propagating in the perpendicular direction. The velocity of P_2 is related to the third eigenvalue of the Christoffel matrix by:

$$\alpha_2 = \sqrt{\lambda_3} = \sqrt{\frac{C}{\rho}}.$$

In the layered structures shown in Figs. 2-3 and 2-4, it is easy to see that $C < A$ and thus that the wave P_2 is slower than P_1.

If we do not take into account the two angles which are necessary to orient the symmetry axis in space, we can see that out of the 5 elastic coefficients which are necessary to describe the elastic properties in the case of hexagonal symmetry, only the 4 coefficients A, C, L, N can be obtained by measuring the seismic velocity of P and S waves along the symmetry axis and perpendicular to it. To get the fifth elastic coefficient F, it is necessary to perform a velocity measurement in one intermediate direction.

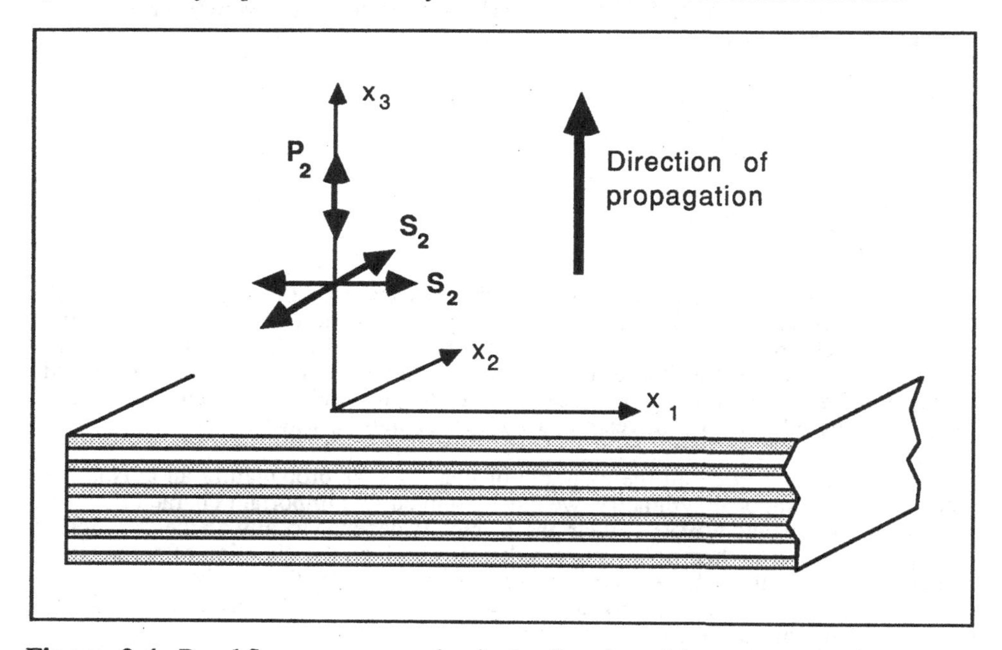

Figure 2-4 P and S waves propagating in the direction of the symmetry axis.

In order to conclude this discussion on hexagonal symmetry systems, let us focus our attention to the important case where the symmetry axis is vertical. Such a case plays an important role in geophysics, in particular if the origin of the seismic anisotropy is directly linked with the gravity field. In seismic exploration, sedimentary sequences with thin horizontal layers of sediments are one example where such a model works. In such structures, the velocities of P waves have values depending on their incidence angle. Waves refracted at a critical angle beneath a horizontal interface propagate at velocity α_1, while vertically reflected or transmitted waves propagate at velocity α_2. These waves are sometimes called P_H and P_V, respectively:

$$\alpha_1 = \sqrt{\frac{A}{\rho}}, \quad \text{velocity of } P_H \text{ waves,}$$

$$\alpha_2 = \sqrt{\frac{C}{\rho}}, \quad \text{velocity of } P_V \text{ waves.}$$

To draw a conclusion from the observation of S waves in such media, it is furthermore necessary to give attention to both their direction of propagation and their polarization. Horizontally propagating S waves have two different velocities depending on their polarization, labelled either S_H, if the polarization vector is horizontal, or S_V if it is vertical. The velocities are:

$$\beta_1 = \sqrt{\frac{N}{\rho}}, \quad \text{for } S_H \text{ waves,}$$

$$\beta_2 = \sqrt{\frac{L}{\rho}}, \quad \text{for } S_V \text{ waves.}$$

In such a transversely isotropic medium with a vertical axis of symmetry, vertically propagating S waves have the same velocity, β_2, as the S_V waves propagating horizontally. Orienting the axis x_3 upward in the previous examples, we have seen that a stack of isotropic layers with alternatively high and low stiffnesses exhibits an overall hexagonal symmetry with A>C and N>L so that $\alpha_1 > \alpha_2$ and $\beta_1 > \beta_2$. This means that long-wavelength S_H and P_H waves are faster than S_V and P_V waves.

By examining in detail the example of plane waves propagating an anisotropic medium with hexagonal symmetry, we have obtained two important characteristics of the anisotropic seismic waves: 1) their phase velocity depends locally on the direction of propagation **n**, and 2) their polarization **a** depends both on the direction of propagation and on the properties of the fourth-order stiffness tensor. Generalization of these properties of plane waves to arbitrary anisotropic homogeneous media is straightforward, as we have seen.

2.3.2 Phase and group velocities

The fact that the phase velocity varies with the direction of propagation leads to a further complication which does not exist in isotropic media. The direction of propagation of the energy of a wave packet may be different from the direction **n** of propagation of the phase: except in particular directions, the group-velocity vector **v** does not coincide with the phase-velocity vector $\mathbf{c} = c\,\mathbf{n}$ in anisotropic media. Actually, the group velocity v may be higher than the phase velocity c, the vector **v** - **c** being perpendicular to **c**. For non-dispersive anisotropic media, such as homogeneous elastic media, Helbig (1958) gives:

$$v_i = c\, n_i + \partial c / \partial n_i \,,$$

where n_i are the Cartesian components of the unit vector **n**. Expressions giving the orientation and modulus of the group velocity vector in terms of the derivatives of the phase velocities with repect to angles of propagation are given by several authors (e.g. Thomsen, 1986, for the hexagonal symmetry case).

The simplest way to visualize this concept of group velocity in non-dispersive media is to consider a plane wave propagating in an anisotropic medium and to place a screen with a large slot parallel to the plane of the wave as shown in Fig. 2-5. If Huygens' principle is used to propagate the plane wave farther by summing up the anisotropic wavelets coming out of the elementary point sources located in the slot and propagating with a phase velocity dependent on the azimuth, Fig. 2-5 shows how constructive interference of the wavelets may take place in an oblique direction with regard to the normal **n** of the incident plane wave.

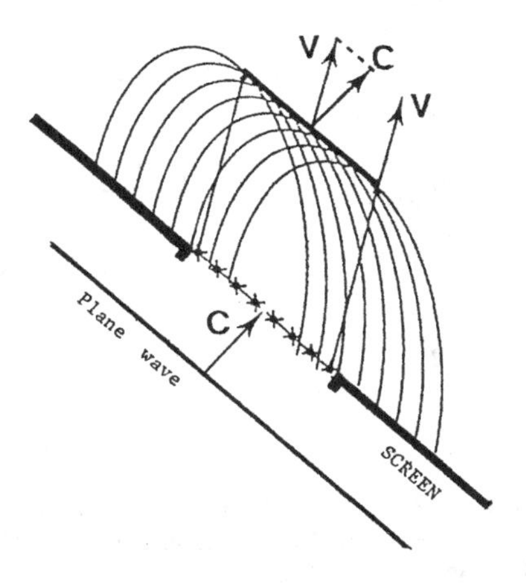

Figure 2-5 Illustration of the concept of group velocity v and phase velocity c in anisotropic media (modified from Garmany, 1989). The phase velocity vector **c** is perpendicular to the phase surface while the group velocity vector **v** is parallel to the direction of the beam of seismic energy.

For real Earth's materials which exhibit weak anisotropy, it seems that the departure of the group-velocity vector from the normal **n** remains quite small. Crampin (1982) has for example established a rule of thumb on upper-mantle minerals, indicating that n% of velocity anisotropy yields between the group and phase velocity vectors an angle smaller than $0.8 \times n$ degrees. Moreover, the difference between the modulus of the phase and group velocities is a second order effect if anisotropy is first order (Backus, 1965). Fig. 2-6 after Crampin (1982) gives the phase and group velocities of olivine and upper mantle anisotropic assemblages.

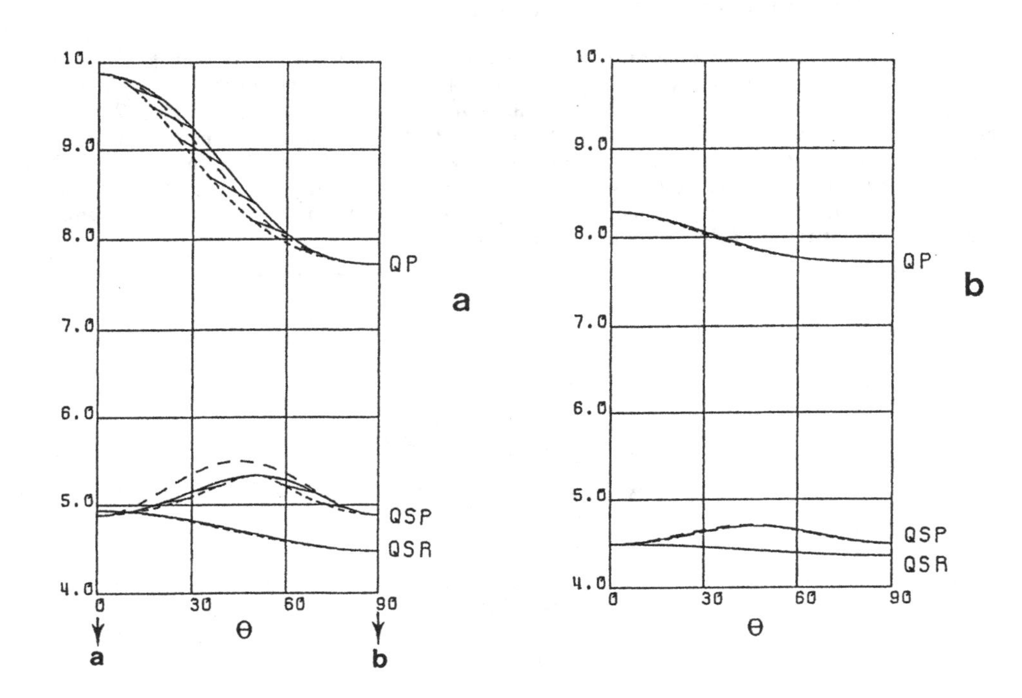

Figure 2-6 Seismic velocities of (a) pure orthorhombic olivine and (b) olivine assemblage made up of 30% oriented olivine crystals with the a-axis (i.e. axis [100]), parallel to the direction $\theta = 0$ and the b- and c-axes (i.e. [010] and [001] axes) randomly oriented in the perpendicular plane. SP and SR represent S waves with polarization parallel and perpendicular to the a-axis respectively. Short dashes correspond to group velocities and solid lines to phase velocities. Long dashes are approximate velocities computed according to Backus (1965), (after Crampin, 1982; AGU, copyright © 1982)

In conclusion to this section devoted to homogeneous anisotropic media, let us mention that severe complications of the seismic wavefield may occur at their boundaries, in particular if two different strongly anisotropic media are in contact. In solid isotropic structures, the transmission and refraction of seismic waves present some complications as compared with acoustics, due to the possibility of conversion from P waves to S_V waves and vice-versa. In strongly anisotropic structures, such wave conversions become more complicated. First, three plane waves polarized at a right angle relative to each other can exist for any propagation direction as we have seen. Second, the polarization of the seismic waves is not simply linked to the propagation direction: S waves are split according to the local symmetry of the elastic tensor. In weakly anisotropic structures, the polarization of the quasi-P wave remains close to the propagation direction and that of quasi-S waves remains close to a perpendicular direction, but contrary to the isotropic case, the quasi-S wave polarizations cannot take an arbitrary orientation in the plane perpendicular to the P-wave polarization.

Reflexion and refraction of plane waves at plane discontinuities between two anisotropic media have been thoroughly studied (see for example Fedorov et al., 1968; Rokhlin et al., 1986; and Li et al., 1990, after Daley and Hron, 1977, 1979). Considering two anisotropic half spaces in a welded contact, the basic equations of continuity for the stress and the displacement fields have to deal with three transmitted waves and three reflected waves for each of the three possible incident waves. Coupling between the polarization of a given wave, its direction of propagation, the local symmetry of the anisotropic material, and the orientation of the discontinuity of the medium have all to be taken into account in computations of the reflection and transmission coefficients of the seismic waves.

2.3.3 Perturbation theory

In weakly anisotropic media, the directions of polarization of the seismic waves remain close to the direction of propagation for the quasi-P waves and to the perpendicular plane for the two quasi-S waves, as we have seen. The perturbation theory can be used both for computing the eigenvector of the Christoffel matrix and its eigenvalues. Backus (1965) has used the first-order perturbation theory to study the angular dependence of the body-wave velocity in such weakly anisotropic media. Developing the square of seismic phase velocity to the first order in perturbation of the stiffness tensor

$$\delta c_{ijkl} = c_{ijkl} - c^0_{ijkl} ,$$

where c^0_{ijkl} denotes a close isotropic tensor. He has obtained a very useful formula giving the angular variation of the quasi-P and -S wave velocities. For the sake of simplicity, we will drop hereafter the term "quasi" to denote P and S waves in media exhibiting a slight departure from isotropy. For P waves, Backus (1965) has found that the angular variation of the phase velocity is given to the first order by a homogeneous trigonometric polynomial of degree 4. Developing this polynomial in Fourier series, he obtained in the plane (x_1,x_2):

$$\alpha(\varphi)^2 - \alpha_0{}^2 = A + C\cos 2\varphi + D\sin 2\varphi + E\cos 4\varphi + F\sin 4\varphi$$

where $\alpha(\varphi)$ and α_0 are the P-wave velocities in the anisotropic and the close isotropic medium respectively, φ is the azimuth of the phase-propagation direction taken from the axis x_1, and A, C, D, E and F depend linearly on the differences δc_{ijkl} between the stiffnesses c_{ijkl} and their isotropic values:

$$A = (3\,\delta c_{1111} + 2\,\delta c_{1122} + 4\,\delta c_{1212} + 3\,\delta c_{2222})\,/\,8\rho,$$
$$C = (\delta c_{1111} - \delta c_{2222})\,/\,2\rho,$$
$$D = (\delta c_{1112} + \delta c_{1222})\,/\,\rho,$$
$$E = (\delta c_{1111} - 2\delta c_{1122} - 4\delta c_{1212} + \delta c_{2222})\,/\,8\rho,$$
$$F = (\delta c_{1112} - \delta c_{1222})\,/\,2\rho.$$

For a practical application of the above relation, the stiffness tensors are commonly computed in a Cartesian coordinate system fixed to a geographical frame with axis x_3 pointing downward and axes x_1 and x_2 being fixed in the horizontal plane (Fig. 2-7), axis x_1 pointing North for example.

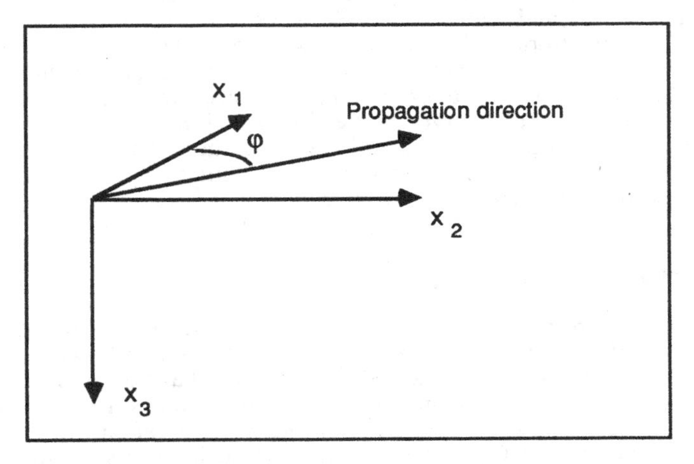

Figure 2-7 Cartesian coordinate system used by Backus (1965) to derive the above expression for angular variation of P-wave velocities in a weakly anisotropic medium.

Crampin (1982) has pointed out that the above relationships, although they are valid in most cases, do not cover all the situations. There are some symmetry systems (triclinic, monoclinic and trigonal) for which higher-order terms, such as the 6φ azimuthal dependence, may become important. It seems, however, that such symmetry systems are unlikely to play an important role in the Earth when overall large-scale anisotropic properties are considered.

Azimuthal variations of S-wave velocities were also given to the first order in anisotropy by Backus (1965). Two S velocities have to be considered. Linear combinations of the squares of these S velocities involve homogeneous trigonometric polynomials of degrees 3 and 6, instead of 4 for P waves.

2.4 Body waves in heterogeneous anisotropic media

All the above remarks pertain to plane waves propagating in homogeneous anisotropic media, whether they are intrinsically anisotropic or equivalent to anisotropic media for long seismic wavelengths. We just mentioned the effect of sharp heterogeneities such as the screen with a slot which led us to introduce the concept of group velocity or planar interfaces between two homogeneous media. We will review now some of the properties of seismic waves in anisotropic media displaying smoothly varying elastic properties. By "smoothly" we mean that the elastic properties only vary by a small fraction over one wavelength of the seismic wave of interest, so that the ray approximation applies to computing the propagation of a seismic wavefront. Developments of the ray theory in anisotropic media were presented by several authors (e.g. Vlaar, 1968; Cerveny, 1972, 1989; Červeny et al., 1977; Gajewski and Psencik, 1987, 1990).

2.4.1 Ray theory

Let us start from the Cauchy equation relating the divergence of the stress field to the acceleration of an elementary volume. Taking the Fourier transform of this equation in Cartesian coordinates and using again the Einstein summation rule, one gets:

$$\partial_j (\hat{\sigma}_{ij}) = - \rho \, \omega^2 \hat{u}_i \, ,$$

where $\hat{u}_i$ and $\hat{\sigma}_{ij}$ denote the time-Fourier transform of u_i and σ_{ij}, ω is the circular frequency. If one substitutes σ_{ij} by its relation given by Hooke's law, one gets the elastodynamic equation of an anisotropic heterogeneous medium:

$$\partial_j (c_{ijkl} \, \hat{\varepsilon}_{kl}) = \partial_j (c_{ijkl} \, \partial_k \hat{u}_l) = - \rho \, \omega^2 \hat{u}_i \, .$$

The basic approach of the ray theory is to consider the high-frequency part of an asymptotic development of the seismic wavefield **u(r**,t) which corresponds to the onset of the particle motion on a wavefront. For this purpose, let us develop the Fourier transform of the seismic wavefield $\hat{\mathbf{u}}$ in the classical ray expansion in inverse powers of ω, so that the high-frequency part of the wavefield corresponds to the low-order terms of the series:

$$\boxed{\hat{u}_i (\omega) = e^{\, i \, \omega \tau(\mathbf{r})} \, \Sigma_n \, (i\omega)^{-n} \, {}_nA_i \, (\mathbf{r}) \, ,}$$

where **r** denotes the space coordinates, $\omega\tau(\mathbf{r})$ is a phase function which gives the rapid oscillation of the seismic wavefield and ${}_nA_i(\mathbf{r})$ is a smoothly varying function in space giving both the amplitude of the wave and its polarization. Introducing this Taylor expansion into the elastodynamic equation and keeping only the zero order term which is the most important one at high frequency, we get:

$$c_{ijkl} \, \omega^2 \, \partial_j (\tau) \, \partial_k(\tau) \, e^{\, i \, \omega\tau} \, {}_0A_l = \rho \, \omega^2 \, e^{\, i \, \omega\tau} \, {}_0A_i \, ,$$

providing $\partial_j \, c_{ijkl} \cdot \partial_k(\tau)$ can be neglected when compared with $c_{ijkl} \, \omega \, \partial_j (\tau) \, \partial_k(\tau)$. It may be useful to comment here on the validity of this approximation which is the basis of the ray theory in seismology. $\partial_j c_{ijkl} \partial_k(\tau)$ can be neglected as compared with $c_{ijkl}\omega\partial_j(\tau)\partial_k(\tau)$ if $\omega\|\mathbf{grad}\tau\|$ is much larger than the relative variations of the elastic coefficients over a unit length. With the introduction of the local phase velocity

$$c(\mathbf{r}) = 1 / \| \mathbf{grad}\ \tau(\mathbf{r}) \| ,$$

this condition means that ω/c (and thus $2\pi/\lambda$ if λ is the wavelength) must be much larger than the relative variations of the elastic coefficients over a unit length. This condition is satisfied if the relative variation of elastic coefficients is much smaller than unity over one wavelength (classical validity condition for the ray theory).

In summary, the ray theory approximation can be applied to the computation of the propagation of the high frequency part of the seismic wavefield in regions where the elastic properties vary smoothly over distances comparable to the wavelength. In cases where abrupt variations of the elastic coefficients would occur, reflections, wave conversions, and possibly diffraction by the discontinuity become dominant in the seismic wavefield behavior. As we have seen above, this may introduce considerable complexities in the propagation of seismic waves in strongly anisotropic media. Even if the situation is more simple in weakly anisotropic media, some serious complications of the wavefield do occur, too, due to shear-wave splitting for example.

Let us now introduce a unit vector $\mathbf{n}$ perpendicular to the surface of constant phase $\omega\tau$:

$$\mathbf{n} = \mathbf{grad}\ \tau / \| \mathbf{grad}\ \tau \| ,$$

The condition to be satisfied by the first term of the ray series expansion of the seismic wavefield is:

$$c_{ijkl}\ n_j\ n_k\ {}_0A_l = \rho\ c^2\ {}_0A_i ,$$

or

$$m_{il}\ {}_0A_l = c^2\ {}_0A_i ,$$

where m_{il} are, as previously, the components of the Christoffel matrix

$$m_{il} = c_{ijkl}\ n_j\ n_k / \rho$$

but which is now space-dependent because c_{ijkl} and ρ depend on the space coordinates.

In order to satisfy the equation of elastodynamics in an anisotropic solid with smoothly-varying elastic properties, we see from the above equations that the square of the local phase velocity $c(\mathbf{r})$ must be an eigenvalue of the local Christoffel matrix $(m_{il}(\mathbf{r}))$ for a given direction $\mathbf{n}$ of the phase propagation which is normal to the curved phase-surface:

$\tau(\mathbf{r})$ = constant.

As in homogeneous models, there are three possible orthogonal directions of polarization, parallel to the eigenvectors of the Christoffel matrix for a given propagation direction **n**. The difference is that the Christoffel matrix depends now on the space coordinates and that we are looking at the local properties of the wavefield, so that both seismic velocities and polarization directions depend on the local properties. When the elastic properties vary smoothly with the space coordinates, the direction of polarization of a given seismic wave may rotate slowly as the wave propagates along a given ray.

In the simple case of isotropic solids, the eigenvalues λ_p (p=1,3) of the Christoffel matrix do not depend on the direction of propagation of the waves, so that one can write:

$\| \mathbf{grad}\, \tau(\mathbf{r}) \|^2 = \lambda_1 = \rho / (\lambda + 2\mu)$ for P waves, and

$\| \mathbf{grad}\, \tau(\mathbf{r}) \|^2 = \lambda_2 = \lambda_3 = \rho / \mu$ for S waves,

λ and μ being the Lamé's coefficients. We then find the classical eikonal equations of isotropic elastic solids for P and S waves. These equations contain all the geometrical properties of the P and S rays along with high frequency waves which can propagate at two independent velocities:

$$\alpha = \sqrt{\frac{\lambda + 2\mu}{\rho}}\ , \quad \text{for P waves, and}$$

$$\beta = \sqrt{\frac{\mu}{\rho}}\ , \quad \text{for S waves.}$$

In an anisotropic solid, on the other hand, the eigenvalues λ_p depend on the direction of propagation **n** of the phase surface. When evaluating the propagation properties of this surface, we have to know the direction of **grad** τ in each point of the surface to determine the phase velocity and to propagate the phase-surface farther.

We have already mentioned that, in general, the group-velocity vector is not parallel to the phase-velocity vector. The seismic energy, and thus the rays which are associated with the propagation of energy, differ from the normal to the phase surface $\tau(\mathbf{r})$=constant (Fig. 2-8). Fortunately, as we have also seen, the differences between both velocities can be considered as a second-order effect in weakly anisotropic media. Taking the departure from isotropy as a small perturbation, Backus (1965) has shown that to the first order in anisotropy, the modulus of the phase and group velocities are the same but that the angles between the phase- and the group-velocity vectors are, on the other hand, of the first order in anisotropy.

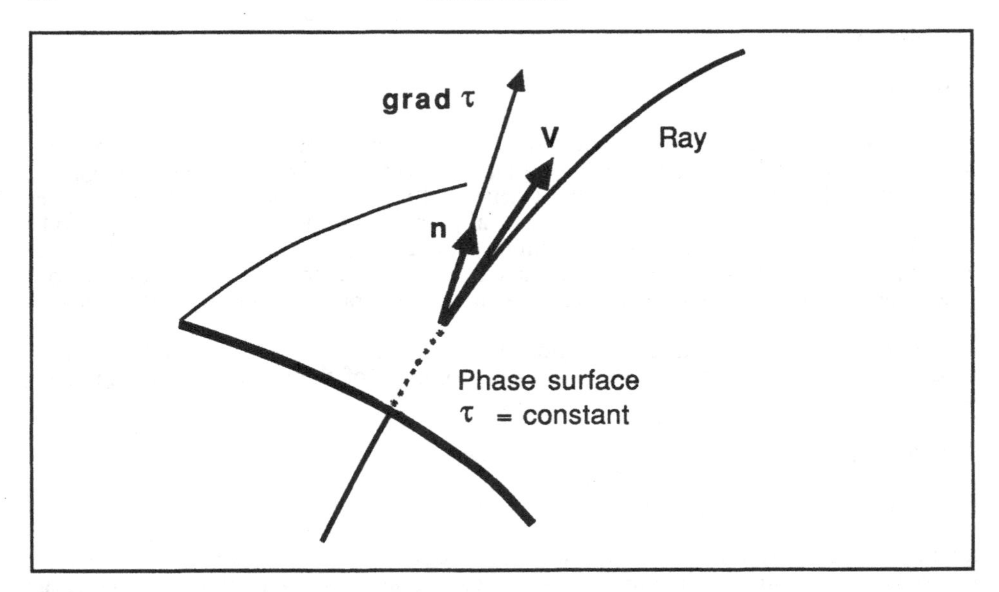

Figure 2-8 Illustration of the eikonal equation in an arbitrary anisotropic medium: the phase surface travels in the direction of **grad**τ at a velocity c(**r**) which is the square root of one of the eigenvalues of the local Christoffel matrix. Seismic energy travels in the direction of the group-velocity vectors **V**, along rays which are not, in general, perpendicular to the phase surface.

2.4.2 Computational problems

Ray theory is the simplest method to compute propagation times and the variations of amplitude of a wave pulse which is due to geometrical spreading in an anisotropic smoothly varying structure. Ray method fails in singular regions such as in the vicinity of a caustic and the boundary of a shadow zone. Various modifications of the ray method may be used to remove these problems (e.g. Cervenyet al., 1977). When waveforms have to be computed in weakly anisotropic and heterogeneous media, ray theory may also be inadequate just because it ignores the possible coupling between the different waves. Each split shear wave, for example, propagates in a way which is completely decoupled from other waves in the ray theory. Coates and Chapman (1990) have shown that this high frequency limit may then lead to misleading results. Coupling between the two split shear waves has to be taken into account to correctly predict the shape of the seismic signal if the time delays between the two shear-wave arrivals are small.

When applying the ray theory to inhomogeneous anisotropic media, numerical computations are much more complicated than in isotropic media since an eigenvector-eigenvalue analysis has to be performed at each step of the propagation along a ray (Garmany, 1983; Frazer and Fryer, 1989). However, even if ray theory is time consuming in anisotropic heterogeneous media, it is much more efficient than computing body wave seismograms with more exact numerical techniques such as finite elements (e.g. Chen, 1984), or the reflectivity method in the case of structures with planar interfaces (Booth and Crampin, 1983; Fryer and Frazer, 1984).

Two basic approaches have been followed in order to reduce the computer time in body-wave computations: perturbation methods (e.g. Farra, 1989, 1990; Jech and Psencik, 1989; Nowack and Psencik, 1991) and factorization of the anisotropic media in homogeneous blocks (Cerveny, 1989; Farra, 1990). Together with various high frequency asymptotic technics (e.g. Garmany, 1988a, 1988b), these approaches presently constitute a very active field of reasearch in theoretical seismology.

2.4.3 Concluding remarks

Let us conclude this section on body waves by saying that anisotropy shows up in two different ways. First of all, phase and group velocities of body waves locally depend on the direction of propagation, whether it is a P wave or one of the two S waves which exist *a priori* in any anisotropic media. The basic difficulty in observing this angular dependence of the seismic velocities is mainly due to the problem of discrimination between the effects of local heterogeneities and the effects of intrinsic anisotropy in the observed travel times. In seismic refraction experiments, there is a simple way to discriminate between a dipping interface and intrinsic anisotropy: Backus (1965) has pointed out that the azimuthal variation of the P-wave velocity due to a dipping interface depends on $\cos\varphi$, while anisotropy produces even order terms in the azimuthal dependence. Effects of curvature of a refracting interface can, on the other

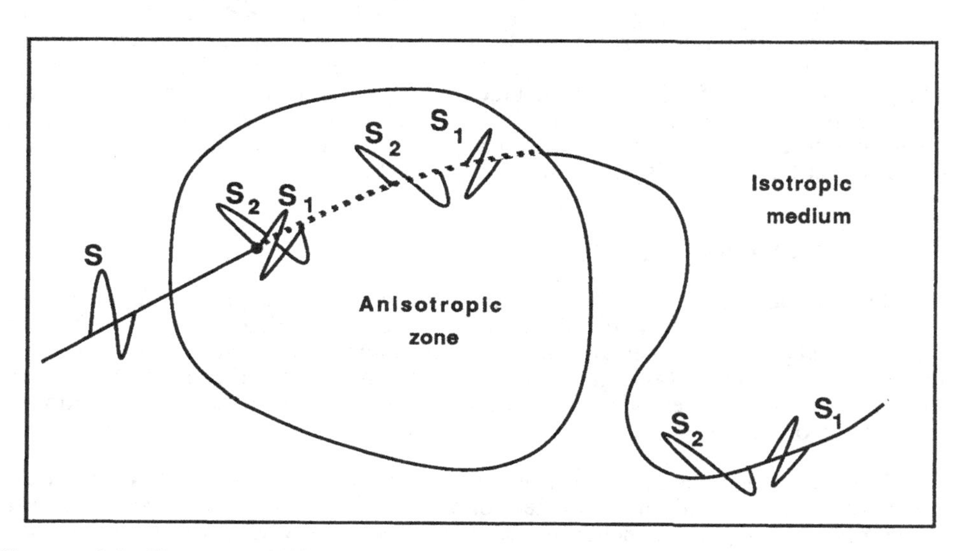

Figure 2-9 Sketch showing S-wave splitting due to the transmission through an anisotropic zone (after Crampin, 1981). An initial S wave split into two waves S_1 and S_2 when entering the anisotropic zone. Ray theory predicts that the two mutually perpendicular polarization vectors can rotate slowly around the ray in a smoothly varying medium.

hand, produce azimuthal variations similar to those connected with anisotropy. Such simple discrimination tests are not so easy to perform in general and the trade-off between large-scale heterogeneities and anisotropy remains one of the most challenging problem of the seismological inverse theory as we will see in Chapter 4.

Another way by which anisotropy can be observed from high-frequency body waves is to look at the S-wave polarization. S-wave polarization is no longer a degenerate property in anisotropic media. When entering such a medium, an S wave with linear polarization transverse to the ray will be split into two S waves with orthogonal directions of polarization in the plane perpendicular to the ray path as we have shown above (Fig. 2-9). In case of a smoothly varying anisotropic medium the polarizations of the two split shear waves may rotate continuously around the ray and the time delay between both waves increases as the waves propagate in the anisotropic medium.

2.5 Surface waves in anisotropic media

Seismic surface waves have proved to be a very powerful tool for investigating the large-scale anisotropic properties of the Earth's upper mantle. The pioneering work of Forsyth (1975) for the Pacific Ocean has revived interest in this subject after some doubts were raised about surface-wave anisotropy in the late 1960's (Boore, 1969; James, 1971).

Surface waves are guided and dispersed waves. Their propagation is parallel to the surface of the Earth. The large wavelengths penetrate deeper than the short wavelengths and in general exhibit greater phase velocities. Along the vertical coordinate, surface waves behave as stationary waves, like stationary vibrations of a finite-length string. The particle motion of the fundamental mode presents no node along the vertical axis. Its velocity for a given period is the lowest. Higher modes display nodes of displacement along the vertical coordinate. For a fixed period they are sensitive to depths larger than the fundamental mode (e.g. Fig. 2-10).

2.5.1 Love and Rayleigh waves

In isotropic laterally homogeneous structures there are two types of surface waves: the Love and the Rayleigh waves. The Love wave particle motion is rectilinear. It lies in the horizontal plane and is perpendicular to the direction of propagation. Love waves are made up of S_H waves. Rayleigh waves are made up of interfering P and S_V waves. Their particle motion is elliptical. It lies in the vertical plane containing the direction of propagation of the waves.

The depth of penetration of a given surface wave thus depends on both the wavelength and the rank of the mode. For a given type of a surface wave and for a given mode, the depth of penetration increases with the wavelength and consequently with the period. Long wavelengths are sensitive to the elastic properties of the deeper layers while short wavelengths are sensitive to superficial layers. For this reason, a given mode of surface waves has different velocities at different wavelengths leading to the dispersion of the seismic signal. The dispersion curves, either the phase or group velocities, are key data for any surface-wave investigation of the Earth's structures. The

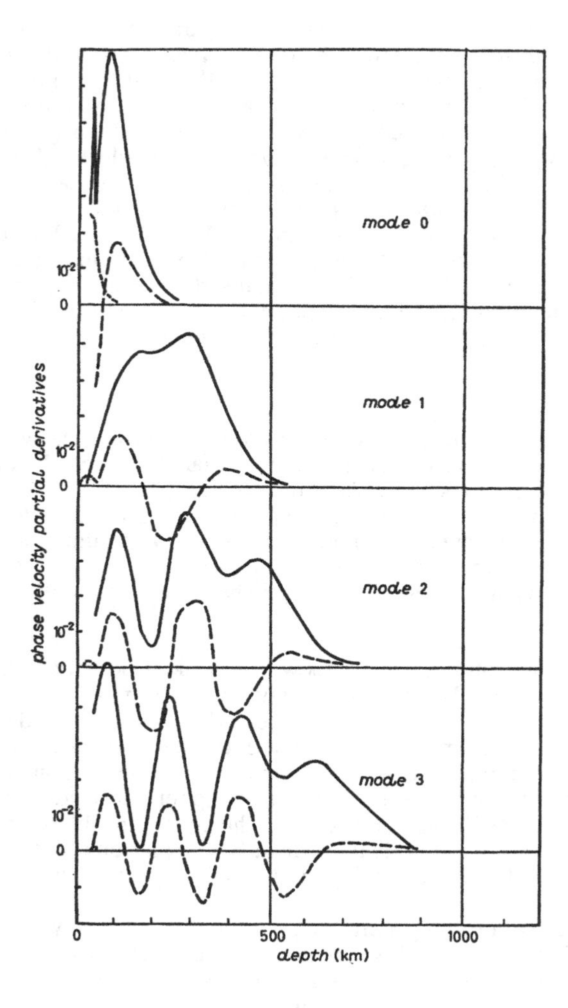

Figure 2-10 Partial derivatives of Rayleigh-wave phase velocity at a fixed period of 50 s with respect to the isotropic S velocity (continuous lines), density (large dashes) and P velocity (small dashes). The curves are drawn for the fundamental mode (mode 0), and overtones 1 to 3 (modes 1 to 3). These partial-derivative curves give the sensitivity of the phase velocity of a given mode to small perturbation of the three physical parameters for 1-km layer thicknesses; the P- and S-velocity partial derivatives are in dimensionless units and the density partial derivatives are in km/s/g/cm^3 (after Cara, 1983).

application of inverse methods is widespread in interpreting these dispersion curves in terms of depth variation of the elastic properties of the Earth.

The wavelengths of the surface waves penetrating into the the upper mantle are much larger than those of the crustal surface waves and mantle body waves. Crustal surface waves have periods which are typically in the range of 10 - 30 s. Their wavelengths reach 30 to 100 km. Mantle surface waves have longer periods and exhibit wavelengths of several hundred kilometers. For periods larger than 200 s, surface waves excited by large earthquakes can circle the whole Earth several times. They penetrate to depths which become a significant ratio of the Earth's radius. To develop the theory of their propagation it is necessary to consider the eigenvibrations of a spherical Earth model. At very long periods, the Love and Rayleigh waves can be represented by a superposition of toroidal and spheroidal vibrations of the whole Earth (e.g. Aki and Richards, 1980). If R denotes the Earth's radius, the relationships between the eigenperiod T_n of a given spherical normal mode of the angular order n and the phase velocity C_n of the associated surface wave is:

$$C_n = 2\pi R / (n+1/2)\, T_n.$$

Because of the very large wavelengths involved in surface-wave investigations, it is clear that only overall anisotropic properties of large units can be investigated with the use of such data. Questions about the physical significance of the observed large-scale anisotropy can only be addressed by taking into account the wavelengths involved. It is also clear that surface waves cannot discriminate between intrinsic anisotropy due, for instance, to the preferred orientation of anisotropic crystals and shape-induced anisotropy caused by small-scale flat isotropic heterogeneities preferentially oriented in a strain field.

In transversely isotropic structures with a vertical axis of symmetry, surface-wave properties are very similar to those obtained in the isotropic case (Harkrider and Anderson, 1962). Such anisotropic structures can be extended with no further complication to spherical models: decoupled toroidal and spheroidal modes still exist, and the same developments as for the isotropic case can be done to represent surface waves as superposition of eigenvibration of spherical modes. Backus (1967) and Takeuchi and Saito (1972) have worked out the general problem of the eigenvibrations of transversely isotropic spherical models. The equations and notations applied by Takeuchi and Saito (1972) are broadly used in the seismological literature.

Surface waves in more general anisotropic media display many peculiar features and are more difficult to analyse. It is even inadequate to use the terms Love and Rayleigh waves when the effects of anisotropy become too strong. The computation of surface-wave propagation in an arbitrary flat-layered structure has been extensively studied by Crampin (1970, 1975) who has generalized the Thomson (1950)-Haskell (1953) technique to take into account the six possible upgoing and downgoing plane waves which can exist in each layer of a stack of layers made of general anisotropic media. Prior to Crampin (1970), different authors discussed the propagation of surface waves in general anisotropic structures (Synge, 1957; Buchwald, 1961). Anisotropic homogeneous half spaces were studied in detail for the cubic symmetry (Stoneley, 1955; Buchwald and Davis, 1963) and the orthorhombic symmetry (Stoneley, 1963).

Following Crampin (1970), a family of generalized surface waves can be defined in arbitrary anisotropic structures. In the case of weak anisotropy, this family splits into two sub-families: the quasi-Love and quasi-Rayleigh waves. These waves have polarizations which are close to the polarizations of the Love and Rayleigh waves in isotropic media. Very roughly speaking, in weakly anisotropic media, azimuthal variations of velocities of the Love waves are similar to those of S_H waves while the azimuthal variations of Rayleigh wave velocities behave more or less like S_V waves propagating horizontally. Since in surface-wave seismology we are mainly concerned with large-scale anisotropic properties of the crust and upper mantle, anisotropic effects are in general small. For this reason, we will continue employing the terms Love and Rayleigh waves, for short, in the following.

2.5.2 Love/Rayleigh-wave incompatibility

Before addressing in greater detail the question of surface-wave propagation in anisotropic media, let us start again with the simplest possible example of an anisotropic structure: the case of transversely isotropic structures with a vertical axis of symmetry. Whether the anisotropy is intrinsic or not, such models are often called transversely isotropic models without referring to the orientation of the symmetry axis because the elastic properties do not depend on the azimuth of propagation of the waves in the horizontal plane. The behavior of Rayleigh waves in such a simple anisotropic model was investigated by Stoneley (1949) for a homogeneous half space. Anderson (1961, 1962) solved the problem for both the Rayleigh and the Love waves in the case of a stack of homogeneous transversely isotropic layers. As there is no azimuthal variation of the elastic properties in such a model, the velocities and the propagation of the surface waves do not depend on the azimuth of propagation. The Love-wave polarization remains of the pure S_H type while the Rayleigh waves remain polarized in the vertical plane containing the direction of propagation. Only the computation of the dispersion curves is affected since five elastic coefficients have to be taken into account.

There are observations of Love- and Rayleigh-wave dispersion curves which cannot be accounted for by simple isotropic models presenting smooth depth variations of velocity. Such observations are often referred to as Love/Rayleigh-wave discrepancies. Transversely isotropic models are well adapted to fit such observations since they allow us to fit simultaneously both sets of dispersion curves without introducing unrealistic velocity fluctuations of the model along the vertical coordinate. Because these observations of Love/Rayleigh-wave discrepancies are directly connected with the polarization of the waves, anisotropy of transversely isotropic structures is sometimes referred to as due to some "polarization anisotropy" in the seismological literature. We will not use such a terminology hereafter because it does not pertain to an unambiguous physical property of the elastic medium. It is directly connected with the observation of an anomaly of the seismic waves and, as it may be due to a different type of anisotropy of the elastic medium, it may create some confusion. Here we have preferred employing systematically the terms directly connected with the physical symmetry of the elastic structure, such as transverse isotropy or hexagonal symmetry with a vertical axis of symmetry in this case.

Anisotropy of the upper mantle was suspected on the basis of several observations of Love/Rayleigh-wave discrepancies in the early sixties, in particular by Anderson and Harkrider (1962), Aki and Kaminuma (1963) and McEvilly (1964). The two latter authors did not compute the dispersion curves for a transversely isotropic body but they rather performed an independent isotropic inversion of the observed Love- and Rayleigh-wave dispersion curves. Although this procedure is not a correct one and can lead to some misinterpretation of the anisotropic properties of the medium (Kirkwood, 1978), it gives us at least some order of magnitude on the S_H/S_V velocity difference which is necessary to fit the data by smoothly varying depth-velocity models. This technique is still quite often used to give a preliminary interpretation of an observed discrepancy between the Love- and Rayleigh-wave dispersion curves (e.g. Journet and Jobert, 1982; Wielandt et al., 1987).

For any type of symmetry of the anisotropic structure other than the one discussed above, azimuthal variations of the surface wave properties appear in addition to a possible Love/Rayleigh-wave discrepancy. Fig. 2-11 adapted after Maupin (1985) shows us, for instance, the azimuthal variations of the Love- and Rayleigh-wave velocities in the case of a simple anisotropic model of the oceanic lithosphere. In this example the upper mantle has been considered as a uniform anisotropic medium with a horizontal axis of symmetry. This model has been designed to mimic the effect of orientation of olivine crystals in the upper mantle. The horizontal symmetry-axis

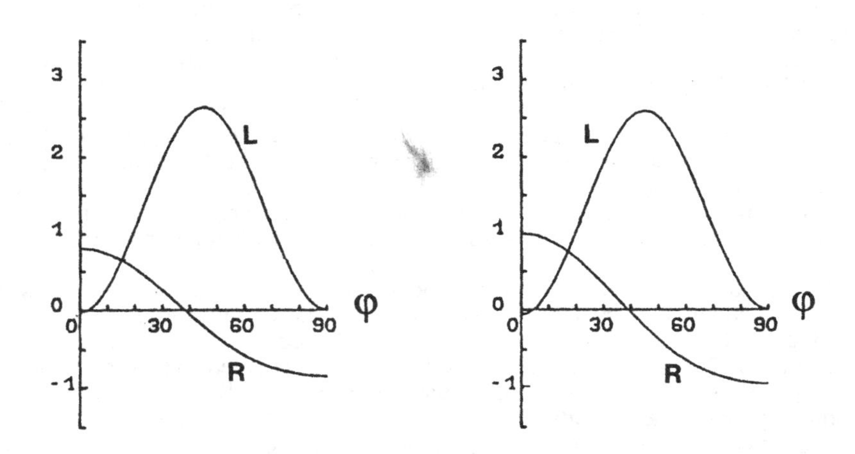

Figure 2-11 Percentage of azimuthal variations of phase velocities for the fundamental modes of Love waves (L) and Rayleigh waves (R) versus azimuth φ in the horizontal plane at two periods: 60 s (left) and 130 s (right). Uniform anisotropy with hexagonal symmetry and a symmetry axis oriented in the horizontal direction $\varphi = 0$ is assumed in the upper mantle. This model mimics the preferred orientation of the a-axis of olivine crystals in the horizontal direction (azimuth zero) while the b- and c-axes are randomly oriented in the perpendicular plane (after Maupin, 1985; reprinted with permission of the Royal Astronomical Society, copyright © 1985).

direction is fast for P waves in this case. The figure shows the phase velocity variations for the fundamental modes of the Love and the Rayleigh waves at two periods. Very similar patterns are observed for the Love waves, on the one hand, and for the Rayleigh waves, on the other. Love waves are fast for a propagation at 45° from the direction of the symmetry axis. Rayleigh waves are fast in the direction of the symmetry axis. A comparison of this figure with Fig. 2-6b shows that the azimuthal variations of the Love- and Rayleigh-wave velocities are very similar to those of S_H and S_V body waves propagating in the horizontal plane (QSP and QSR in the notations of Fig. 2-6). It is also worth noticing that the azimuthal variation of the Rayleigh-wave velocity exhibits the same trend as that of the P waves in such a model.

2.5.3 Perturbation theory

In the case of small anisotropy, the perturbation theory can be used for surface waves as well as for body waves. The basic idea is to consider the anisotropic elastic tensor as a sum of an isotropic elastic tensor plus a small anisotropic component as was done by Backus (1965) for body waves. Complications due to the possibility of mode coupling effects can occur if the mode under consideration is too close to other modes in the wavenumber-frequency plane (Maupin,1989). In this case, the use of the quasi-degenerate perturbation theory becomes necessary since one cannot consider a given mode separately from the other close modes. There are cases where such a situation does not occur, at a period longer than 30 s for the fundamental Rayleigh mode for instance, where the phase-velocity curve is far away from other dispersion curves. The use of the ordinary perturbation theory is then fully justified. Within the framework of the non-degenerate perturbation theory, Smith and Dalhen (1973) have derived the following relation for the azimuthal phase-velocity variations of the Love and Rayleigh waves at a given period:

$$C(\varphi) = A_1 + A_2 \cos 2\varphi + A_3 \sin 2\varphi + A_4 \cos 4\varphi + A_5 \sin 4\varphi \,,$$

where the coefficient A_n depends on the elastic properties of the medium and on the period. In this expression, φ is the azimuth of the propagation direction taken positively from North to East. For the specific case of an elastic medium with hexagonal symmetry and a horizontal axis of symmetry, Maupin (1985) noticed that the coefficient A_n could be related to the partial derivative of a transversely isotropic model with a vertical axis of symmetry. In a more general approach, Montagner and Nataf (1986) found that the coefficients A_n depended on depth integrals of 13 linear combinations of the elastic coefficients. More precisely, stating

$$C(\varphi) = C_0 + \delta C(\varphi),$$

C_0 being the phase velocity for the starting isotropic model, they found that $\delta C(\varphi)$ was expandable in a trigonometric relation similar to that of Smith and Dalhen (1973) where the transversely isotropic terms are related to depth integrals of the following linear combinations of elastic coefficients, written in the matrix form introduced in section 2.2:

$$A = 3\,(C_{11} + C_{22})/8 + C_{12}/4 + C_{66}/2,$$
$$C = C_{33},$$
$$F = (\,C_{13} + C_{23}\,)\,/\,2,$$
$$L = (\,C_{44} + C_{55}\,)\,/\,2,$$

$N = (C_{11} + C_{22})/8 - C_{12}/4 + C_{66}/2.$

The cos2φ and sin2φ terms are related to depth integrals of the following terms:

$$G_c = (C_{55} - C_{44})/2,$$
$$G_s = C_{54},$$
$$B_c = (C_{11} - C_{22})/2,$$
$$B_s = C_{16} + C_{26},$$
$$H_c = (C_{13} - C_{23})/2,$$
$$H_s = C_{36},$$

where the lower index c refers to the cosine terms, and s to the sine terms. Finally, the cos4φ and sin4φ terms depend on depth integrals of the following coefficients:

$$E_c = (C_{11} + C_{22})/8 - C_{12}/4 - C_{66}/2,$$
$$E_s = (C_{16} - C_{26})/2.$$

In the above relations, C_{ij} refers to the usual matrix notation for stiffnesses computed in a Cartesian coordinate system with axis x_3 oriented downward and azimuth φ counted positively from axis x_1 as in Fig. 2-8.

The above coefficients do not appear in the same way in the Love and Rayleigh wave relations. The azimuthal variations of the phase velocity of Rayleigh waves $\delta C(\varphi)$ depend on all the above coefficients except N, while for Love waves $\delta C(\varphi)$ depends only on N, L, G_c, G_s, E_c and E_s. If we compute the partial derivative of the 2φ or 4φ terms of the trigonometric expansion of $\delta C(\varphi)$ with respect to the parameters B_c, B_s, E_c, E_s, G_c, G_s, H_c and H_s, the advantage is that the depth-integral relations turn out to be identical to those obtained when we derived the isotropic part of $\delta C(\varphi)$ with respect to the five Love coefficients of a transversely isotropic structure A, C, F, L, N (Montagner and Nataf, 1986). This result is of considerable practical interest since it allows us to use the partial derivatives of the phase velocities computed either in a flat Earth structure or a spherical Earth structure using the theory for a transversely isotropic model (Takeuchi and Saito, 1972). These results constitute the basis of the vectorial tomography method developed for surface waves by Montagner and Nataf (1988).

The other important application of the above relation is that it puts on a clear theoretical basis the distinction between the transversely isotropic part of the surface-wave velocities with the zero-φ terms and their azimuthal variations (azimuthal anisotropy). Indeed, if the surface-wave velocities are averaged with the azimuth in a laterally homogeneous region, the resulting average dispersion curves can exhibit a Love /Rayleigh-wave discrepancy which can be interpreted in terms of depth integrals of the five coefficients A,C, F, L, N, even if the structure does not exhibit a symmetry around the vertical axis.

2.5.4 Concluding remarks

Observations of Love-Rayleigh-wave discrepancies and azimuthal variations of phase velocities are the key data in surface-wave seismology. They allow us to put some constraints on the anisotropic models of the Earth. A third type of surface-wave observations might also provide us with some information on anisotropy. As noted

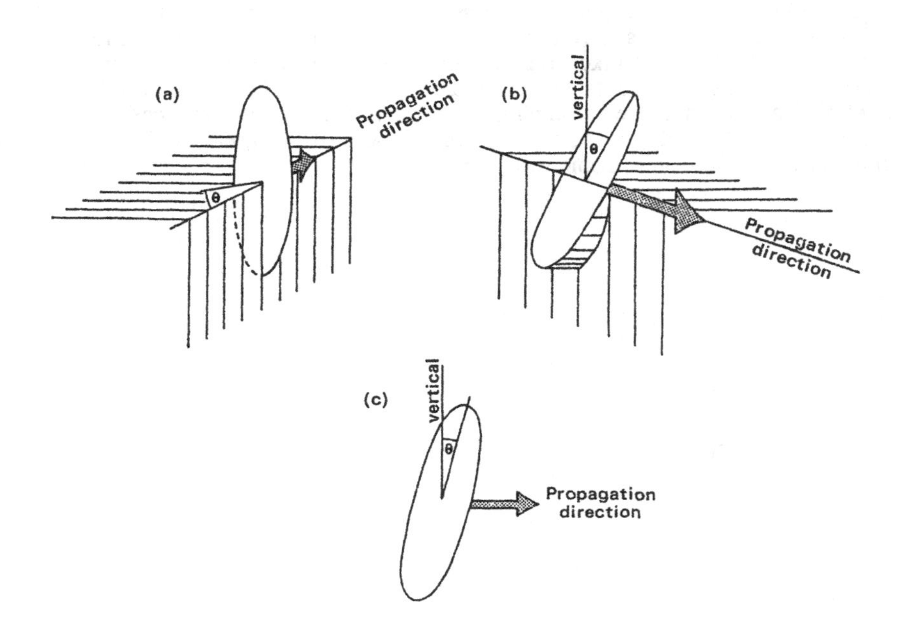

Figure 2-12 Three types of anomalous particle motion of Rayleigh waves which are possible in general anisotropic structures (after Crampin, 1975). In isotropic structures, the Rayleigh-wave particle motion is an ellipse lying in the vertical plane containing the propagation direction. (With permission of the Royal Astronomical Society, copyright © 1975).

above, except when the anisotropic medium displays a vertical axis of symmetry, anisotropy is responsible for several anomalies in the polarization of the surface-wave particle motion (e.g. Crampin, 1975 and Fig. 2-12). Strong anomalies in the polarization of both the Love- and Rayleigh-wave particle motion are quite frequently observed (e.g. Crampin and King, 1977; Kirkwood and Crampin, 1981; Neunhöfer and Malichewsky, 1981). This is in particular the case in a period as short as 10 to 20 s where both elliptical motions of Love waves and tilted and inclined ellipses of Rayleigh waves are commonly observed. The interpretation of these anomalies in terms of anisotropy is by no means unique since a tilted interface connected with strong lateral heterogeneities at the proximity of the seismic station, could also produce anomalies in the particle motion of surface waves (Maupin, 1987).

Finally, to conclude this discussion on surface-wave behavior in anisotropic structures, let us mention that the theory of surface-wave propagation in a spherical fully-anisotropic model requires that a spherical harmonic expansion of the fourth-order elasticity tensor has to be performed. The development of the theory for a fully anisotropic spherical model is by far not obvious. It has been worked out by Mochizuki (1986) making use of the generalized spherical harmonics of Gelfand and Shapiro (1956) (see also Phinney and Burridge, 1973). Splitting of the modes due to anisotropy is one phenomenon to be taken into account for lateral heterogeneities in a spherical model. In a homogeneous anisotropic sphere, for example, the period of the vibration of the whole Earth as measured along a great circle path from interferences of long period surface waves, depends on the location of the pole of the great circle relative to the symmetry axes of the anisotropic medium.

Chapter 3 - Elastic anisotropy of rock-forming minerals and polycrystalline rocks

The velocities of elastic waves in minerals and rocks, and their anisotropy, established in carefully controlled laboratory experiments which simulate the physical conditions in the Earth's interior, provide the necessary basis for realistic interpretations of seismological observations. Velocities of elastic waves are measured in single crystals, naturally occurring rocks and in hot-pressed polycrystalline aggregates. The velocities in rocks and hot-pressed aggregates are commonly affected by cracks and pores which have to be closed by the application of a high confining pressure. Velocities in single crystals are needed for interpretation of the seismic anisotropy resulting from preferred orientation of minerals in rocks.

3.1 Methods of investigation

The experimental methods of measuring the elastic parameters of minerals and rocks are either static or dynamic. The static methods are performed under isothermal conditions and are based on observations of static deformations (bending, twisting, compressing) of oriented crystals which mostly have the shape of a bar or a plate. The number of measurements needed to determine the elastic constants depends on the symmetry of crystals. The accuracy of the static methods is limited by the accuracy of determining small deformations (optical, electrical, mechanical methods) and by the size of the specimens. At present, static methods are mainly used for the investigation of metals, glasses and construction materials which provide homogeneous samples of a large size.

The dynamic methods make use of the vibration properties of crystals. They constitute the main source of information for seismological interpretation. Due to the very high frequencies, the process itself as well as the elastic constants obtained, can be treated as adiabatic. Ide (1935) introduced the dynamic method of resonance vibrations of cylindrical specimens. The principal modes of vibrations are extensional, flexural and torsional. Sophisticated techniques can be applied to determine the elastic properties of spherical or parallel-sided specimens as small as 1 mm (Goto et al., 1976). Furthermore, only one piece of a properly shaped specimen is sufficient to determine all the elastic parameters, because complete information is included in the resonance spectrum of a specimen (Ohno, 1976). The method is also very useful for determining elastic constants at high temperatures (Goto et al., 1989; Isaak et al., 1989). Brillouin scattering measurements yield elastic properties of crystals even smaller than 0.1 mm (Weidner, 1975; Furnish and Basset, 1983) and can be used for measurements at very high pressures.

While dynamic measurements at different frequencies usually give consistent values of elastic constants (Gebrande, 1982), large differences are sometimes observed between static and dynamic measurements, especially at atmospheric pressure. It was shown by Simmons (1964) and Brace (1965) for granite, quartzite and diabase that static compressibilities were systematically greater than dynamic values. At elevated pressures, however, the differences are within the range of accuracy.

The pulse-transmission technique and the resonance methods are probably the most widely used methods of determining elastic parameters in specimens of minerals and rocks at high pressures and temperatures. By the pulse transmission method the transit time of a high-frequency pulse through a specimen is measured in special directions. The pulses of piezoelectric frequencies from 50 kHz to 10 MHz (Anderson and Liebermann, 1966) are generated by crystal or ceramic transducers attached to the specimen. The elastic properties of rocks can be determined by this technique for wavelengths about three times greater than the diameter of mineral grains and size of samples equal to at least five wavelengths (Anderson and Liebermann, 1966).

The elastic parameters have often been considered as practically frequency independent. If it were the case, the velocities of ultrasonic waves in laboratory rock samples could be considered equal to the velocities of seismic body waves. But it is well known that dispersion accompanies intrinsic absorption and, therefore, elastic moduli depend on frequency in a real inelastic material. In general, velocities increase with increasing frequency, which means a systematic velocity decrease from laboratory data to field observations and from body waves to surface waves. Corrections to take into account the physical dispersion in the mantle were proposed by Kanamori and Anderson (1977) and are now commonly used although the physical mechanisms responsible for this attenuation have not yet been clearly identified (Karato and Spetzler, 1990). Intrinsic absorbtion depends on the defects of minerals while elastic properties depend on the geometry of the crystalline lattice. Should physical dispersion of seismic velocity be a source of problems in comparison of high-frequency measurements with observations at seismic frequencies, its effects on anisotropy is likely to be negligible.

The velocities of P waves in rocks, which are determined by pulse transmission techniques, have mostly been studied on cylindrical specimens cored in three mutually perpendicular directions and only exceptionally in more directions (Christensen and Ramananantoandro, 1971; Babuska, 1972b). The accuracy of determining elastic anisotropy from three perpendicular directions depends mainly on the choice of suitable directions of the measurement relative to the velocity extremes in rocks and their fabric. This problem has been solved by measuring the elastic anisotropy on spherical rock specimens at atmospheric pressure (Pros and Babuska, 1967; Thill et al., 1969), or at high hydrostatic pressure (Pros and Podrouzkova, 1974). This enables us to investigate the spatial distribution of the velocities in many directions, to determine the real value of the anisotropy coefficient and to correlate the elastic anisotropy with the rock fabric.

The magnitude of elastic anisotropy may be represented by the coefficient of anisotropy $k = [(v_{max} - v_{min})/ \text{mean } v] . 100\%$ (Birch, 1960a). Several authors used a modified formula to calculate the anisotropy coefficients (Bayuk et al., 1982; Ji and Mainprice, 1988). Therefore, their coefficients were recalculated in accordance with Birch's definition of k, before being applied in this section.

Besides the direct laboratory measurements, the elastic anisotropy of polycrystalline rocks can be calculated if we know the volume fraction, density, elastic constants C_{ij} and the lattice preferred orientation of each mineral (Mainprice, 1990). The calculation can be made at any temperature and pressure if the appropriate elastic derivatives are known. The volume fraction can be determined by the classical point counting technique in thin sections of rocks or by a computerized method (Allard and

Sotin, 1989). The preferred orientation of minerals is usually determined by measurements of individual grains with an optical polarization microscope equipped with a universal stage (for review see Wenk, 1985).

3.2 Velocity anisotropy in rock-forming minerals

The measurements of elastic parameters are the first step in predicting and interpreting seismic velocities. The upper 100 kilometers of the Earth are relatively well known through direct sampling of overthrust pieces of mantle or from xenoliths. At a larger depth, only indirect information based on laboratory experiments is available (high pressure phases). In general, the elastic anisotropy of minerals is closely linked with their crystal symmetry. Elastic properties of minerals depend on both the crystal structure and the chemical composition.

The elastic parameters of the principal rock-forming minerals - the maximum and the minimum velocities of compressional and shear waves, as well as the coefficients of anisotropy k are shown in Table 3-1. The corresponding directions of wave propagation and the polarization directions are given by Miller indices; and those directions which cannot be defined by simple indices are determined by the angles φ and θ (Fig. 3-1). The original data depend on the quality of the investigated crystals and the method

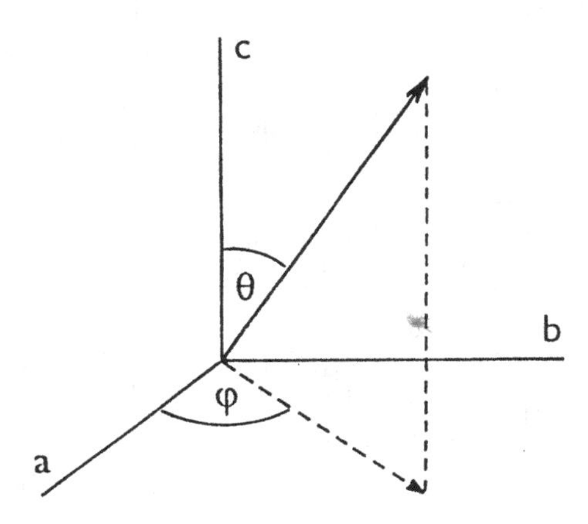

Figure 3-1 Angles defining the propagation and the polarization directions given in Table 3-1.

Table 3-1 Elastic-wave velocities (v_P, v_S) and coefficients of elastic anisotropy (k) in rock-forming minerals.

Mineral	Symmetry	Density ($g \cdot cm^{-3}$)	Compressional waves					
			v_P^{max} (km/s)	Propagation direction	v_P^{min} (km/s)	Propagation direction	v_P^{mean} (km/s)	k%
Ortho- and Ring Silicates								
Olivine (Fo 92. 72)	Orthorhombic	3.31	9.89	[100]	7.72	[010]	8.81	24.6
Zircon	Tetragonal	4.70	9.00	[001]	7.41	[010]	8.21	19.4
Garnets								
Pyrope (PY-1)	Cubic	3.705	8.90	[001]	8.85	$\vartheta=55°$ $\varphi=45°$	8.88	0.6
Almandine (AL-4)	Cubic	3.930	8.76	[001]	8.73	$\vartheta=55°$ $\varphi=45°$	8.75	0.3
Spessartine (SP-1)	Cubic	4.172	8.55	[001]	8.54	$\vartheta=55°$ $\varphi=45°$	8.55	0.1
Grossular (GR-1)	Cubic	3.659	8.95	[001]	8.74	$\vartheta=55°$ $\varphi=45°$	8.85	2.4
Kyanite	Triclinic	3.62	9.88	[001]	9.00	[100]	9.44	9.3
Topaz	Orthorhombic	3.52	10.10	$\vartheta=90°$ $\varphi=56.25°$	9.15	[001]	9.63	9.9
Staurolite	Monoclinic	3.369	10.09	[100]	6.61	[001]	8.35	41.7
Epidote	Monoclinic	3.40	8.38	[010]	7.24	[011]	7.81	14.6
Beryl	Hexagonal	2.64	10.61	[100]	9.76	$\vartheta=36°$ $\varphi=0°$	10.19	8.3
Tourmaline	Trigonal	3.10	9.38	$\vartheta=94°$ $\varphi=33.75°$	7.29	[001]	8.34	25.1
Chain Silicates								
Pyroxenes								
Bronzite	Orthorhombic	3.34	8.30	[100]	7.04	[010]	7.67	16.4
Diopside	Monoclinic	3.31	8.60	[001]	6.94	[101]	7.77	21.4
Hedenbergite	Monoclinic	3.41	6.55	[001]	6.26	[010]	6.41	4.5
Augite	Monoclinic	3.32	8.36	[101]	6.81	[010]	7.59	20.4
Aegirine	Monoclinic	3.50	8.21	[001]	6.75	[101]	7.48	19.5
Hornblende	Monoclinic	3.15	8.13	[001]	6.18	[101]	7.16	27.2
Sheet Silicates								
Muscovite	Monoclinic	2.79	8.06	[110]	4.44	[001]	6.25	57.9
Biotite	Monoclinic	2.89	7.80	[010]	4.21	[001]	6.01	59.7
Framework Silicates								
Feldspars								
Microcline	Triclinic	2.56	8.15	[010]	5.10	[100]	6.63	46.0
Orthoclase	Monoclinic	2.54	7.64	[010]	4.76	[101]	6.20	46.4
Albite (An 9)	Triclinic	2.61	7.26	[010]	5.31	[101]	6.29	31.0
Oligoclase (An 29)	Triclinic	2.64	7.55	[010]	5.50	[101]	6.53	31.4
Labradorite (An 53]	Triclinic	2.68	7.80	[010]	6.06	[100]	6.93	25.1
Anorthite	Triclinic	2.76	8.61	[010]	6.01	[101]	7.31	35.6
Quartz	Trigonal	2.66	7.00	$\vartheta=130°$ $\varphi=90°$	5.36	$\vartheta=72°$ $\varphi=90°$	6.18	26.5
Nepheline	Hexagonal	2.62	7.12	[001]	5.61	[100]	6.37	23.7
Non-Silicates								
Rutile	Tetragonal	4.264	10.65	[001]	8.00	[100]	9.33	28.4
Spinel ($MgO \cdot 3.5\ Al_2O_3$)	Cubic	3.63	10.31	[101]	9.10	[001]	9.71	12.5
Magnetite	Cubic	5.18	7.46	$\vartheta=55°$ $\varphi=45°$	7.29	[001]	7.38	2.3
Pyrite (2)	Cubic	5.013	8.40	[001]	7.78	[110]	8.09	7.7
Calcite	Trigonal	2.717	7.55	$\vartheta=70°$ $\varphi=90°$	5.43	[001]	6.49	32.7

Shear waves								Source of data
v_S^{max} (km/s)	Propagation direction	Polarization direction	v_S^{min} (km/s)	Propagation direction	Polarization direction	v_S^{mean} (km/s)	k%	
5.53	$\vartheta=45^\circ$ $\varphi=0^\circ$	$\vartheta=135^\circ$ $\varphi=0^\circ$	4.42	[010]	[100]	4.98	22.3	Kumazawa and Anderson 1969
4.87	[010]	[001]	2.94	[110]	[110]	3.91	49.4	Ryzhova et al. 1966
4.96	[101]	[10$\bar{1}$]	4.91	[001]	[110]	4.94	1	
4.96	[110]	[1$\bar{1}$0]	4.91	[001]	[1$\bar{1}$0]	4.94	1	Babuška et al. 1978a
4.82	[110]	[1$\bar{1}$0]	4.80	[001]	[110]	4.81	0.4	
5.45	[101]	[10$\bar{1}$]	5.20	[001]	[110]	5.33	4.7	
6.76	[010]	[001]	4.24	[001]	[100]	5.50	45.8	Belikov et al. 1970
6.15	[001]	[100]	5.16	[$\bar{1}$10]	$\vartheta=90^\circ$ $\varphi=41.25^\circ$	5.66	17.5	Voigt 1928
5.29	$\vartheta=118^\circ$ $\varphi=45^\circ$	$\vartheta=62^\circ$ $\varphi=315^\circ$	3.23	$\vartheta=140^\circ$ $\varphi=45^\circ$	$\vartheta=50^\circ$ $\varphi=86.7^\circ$	4.26	48.4	Aleksandrov and Ryzhova 1961a
5.11	[011]	[011]	3.39	[001]	[010]	4.25	40.5	Ryzhova et al. 1966
6.25	$\vartheta=44^\circ$ $\varphi=0^\circ$	$\vartheta=136.7^\circ$ $\varphi=0^\circ$	5.34	[001]	[110]	5.80	15.7	Hearmon 1956
6.14	$\vartheta=82^\circ$ $\varphi=33.75^\circ$	$\vartheta=88.5^\circ$ $\varphi=124^\circ$	4.55	$\vartheta=172^\circ$ $\varphi=33.75^\circ$	$\vartheta 389.4^\circ$ $\varphi=119.4^\circ$	5.35	29.7	
4.99	[010]	[001]	4.27	$\vartheta=90^\circ$ $\varphi=45^\circ$	$\vartheta=90^\circ$ $\varphi=135^\circ$	4.63	15.6	Kumazawa 1969
4.83	[011]	[0$\bar{1}$1]	3.94	[110]	[1$\bar{1}$0]	4.39	20.3	Aleksandrov et al. 1963 Volarovich et al., 1975
4.72	[011]	[0$\bar{1}$1]	3.81	[100]	[001]	4.27	21.3	Aleksandrov et al. 1963
4.65	[011]	[0$\bar{1}$1]	3.48	[010]	[100]	4.07	28.7	
4.60	[011]	[0$\bar{1}$1]	3.37	[001]	[100]	3.99	30.8	Aleksandrov and Ryzhova 1961b
5.01	[010]	[100]	2.03	[001]	[100]	3.52	84.7	Aleksandrov and Ryzhova 1961c
5.06	[010]	[100]	1.34	[010]	[001]	3.20	116.3	Belikov et al. 1970
4.96	[011]	[01$\bar{1}$]	2.14	[010]	[001]	3.55	79.4	Aleksandrov and Ryzhova 1962
4.45	[011]	[01$\bar{1}$]	2.33	[001]	[010]	3.39	62.5	Ryzhova and Aleksandrov 1965
4.63	[011]	[01$\bar{1}$]	2.59	[001]	[010]	3.61	56.5	
4.70	[011]	[01$\bar{1}$]	2.65	[001]	[010]	3.68	55.7	Ryzhova 1964
4.76	[011]	[01$\bar{1}$]	2.70	[001]	[010]	3.73	55.2	
4.96	[011]	[01$\bar{1}$]	2.91	[010]	[100]	3.94	52.0	Aleksandrov et al. 1974
5.06	[100]	$\vartheta=147^\circ$ $\varphi=284.5^\circ$	3.35	[100]	$\vartheta=122^\circ$ $\varphi=90^\circ$	4.21	40.1	Huntington 1958
3.92	[011]	[01$\bar{1}$]	2.83	[100]	[010]	3.38	32.2	Ryzhova and Aleksandrov 1962
6.74	[100]	[010]	3.32	[110]	[110]	5.03	68.0	Birch 1960b
6.61	[001]	[1$\bar{1}$0]	4.50	[101]	[10$\bar{1}$]	5.56	37.9	Verma 1960
4.29	[001]	[1$\bar{1}$0]	4.06	[101]	[10$\bar{1}$]	4.18	5.5	Hearmon 1956
5.66	[110]	[$\bar{1}\bar{1}$0]	4.70	[001]	[110]	5.18	18.5	Aleksandrov and Ryzhova 1961d
4.77	$\vartheta=126^\circ$ $\varphi=90^\circ$	[100]	2.66	$\vartheta=36^\circ$ $\varphi=90^\circ$	[100]	3.72	56.7	Hearmon 1956

of measurement. An accurate determination of the elastic parameters requires gem quality crystals of a reasonable size. This is one of the reasons why the elastic constants of single crystals are missing for several minerals such as chlorite and serpentine.

The minerals in Table 3-1 are arranged from ortho- and ring silicates, via chain silicates, sheet silicates, and framework silicates to non-silicates, as is usual in systematic mineralogy. Of the group of ortho- and ring silicates, olivine and garnets are among the most abundant rock-forming minerals in the lithosphere. Olivine (orthorhombic symmetry, 9 elastic constants) is an essential component in most groups of ultramafic xenoliths and meteorites. For these reasons it is generally considered to be a major part of the upper mantle. The large anisotropy of olivine (25% and 22% for P and S velocities, respectively) is a favorable condition for the world-wide anisotropy observed in the upper mantle. As a result of the frequent and well developed preferred orientation of olivine, olivine-rich ultramafic rocks usually have a high degree of velocity anisotropy, i.e. about 10% for the compressional waves on average (Babuska, 1984). In contrast to olivine, garnets (cubic symmetry, 3 elastic constants), show the lowest anisotropy of all the minerals listed. An admixture of garnets in the ultramafic rocks tends to increase the average velocities and to decrease their elastic anisotropy.

In the important group of chain silicates, the published data allow the elastic anisotropies to be calculated for only a few pyroxenes and one amphibole (hornblende). Orthopyroxene (bronzite, orthorhombic symmetry, 9 elastic constants) displays an anisotropy of 16% for both the P and S wave velocities. However, in olivine-orthopyroxene aggregates the maximum P velocity in orthopyroxene (along the a-axis, Table 3-1) parallels the minimum (b-axis) velocity of olivine (Christensen and Salisbury, 1979), and thus the orientation of orthopyroxene grains in ultramafic rocks decreases the overall anisotropy. The high elastic anisotropy of hornblende and its preferred orientation result in a high anisotropy of amphibolites (Siegesmund et al.,1989). As regards sheet silicates, the elastic constants of single crystals can be determined reliably only for the main members of the mica group. The extremely high values of the coefficients of anisotropy in micas are explained by their structure in which a layer of octahedrally coordinated cations is sandwiched between two identical layers of linked $SiAlO_4$ tetrahedra (Deer et al., 1966). A remarkable anisotropy of the physical properties, including a perfect basal cleavage, is a consequence of the great difference in the strength of interatomic bonds in the directions parallel and perpendicular to the layered atomic structure. This is particularly important in highly deformed crustal rocks such as schists and mylonites. The perfect cleavage, however, may decrease the accuracy of determining the elastic parameters and may cause a considerable scatter of data.

Framework silicates include the most abundant minerals of the Earth's crust - feldspars (triclinic symmetry, 21 constants) and quartz (trigonal symmetry, 7 constants). Among feldspars the potassium members belong, along with the mica group, to the most anisotropic minerals. In the plagioclase feldspar series the coefficients of S-wave anisotropy are substantially higher than those for for the P waves. The elastic anisotropy of quartz, though lower in comparison with feldspars, still classes it with the rock-forming minerals with high coefficients of anisotropy. This is in contrast to the generally very low coefficients of the crack-free elastic anisotropy of the granitic rocks (Birch, 1961; Babuska et al., 1977).

Among non-silicates probably only calcite, in some regions of the uppermost crust, and spinel in the spinel peridotites, may sporadically form substantial volumetric parts of the lithosphere. The high elastic anisotropy of calcite, together with its preferred orientation, results in a high velocity anisotropy of marbles (Babuska, 1968; Kern, 1974). Due to a smaller P-wave anisotropy of spinel (cubic symmetry, 3 constants), any admixture of this mineral in peridotite or lherzolite would decrease their anisotropy of compressional velocities. As the content of spinel in ultramafic rocks is usually very small (up to 6%), its diluting effect on the overall anisotropy is practically negligible.

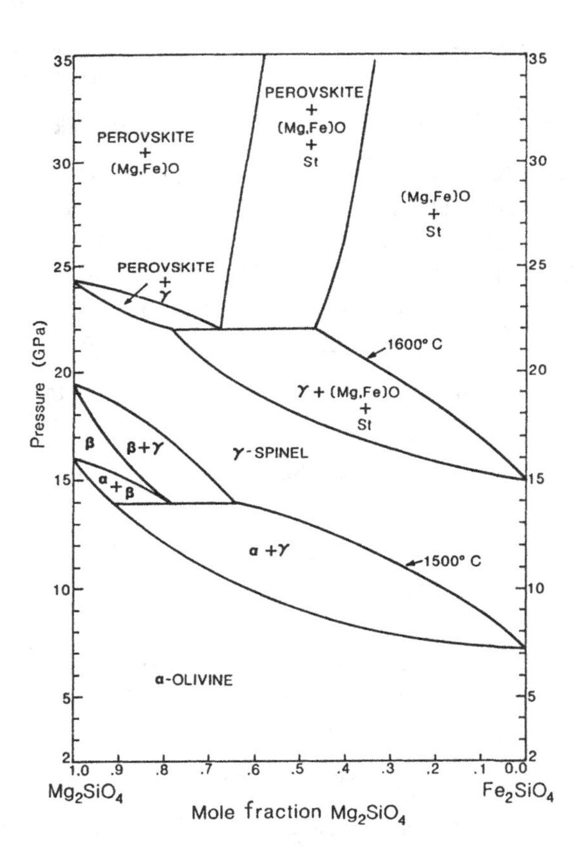

Figure 3-2 Phase diagram for high pressure phases of olivine (after Anderson, 1989). (With permission of Blackwell Scientific Publications, copyright © 1989).

Table 3-2 Stiffness tensor components of the high pressure phase of upper mantle minerals given in the matrix notation (C_{ij}) of Chapter 2 (*). Units are in GPa.

β-Spinel (γ - Mg_2SiO_4) (1)

$$(C_{ij}) = \begin{pmatrix} 360 & 75 & 110 & 0 & 0 & 0 \\ 75 & 383 & 105 & 0 & 0 & 0 \\ 110 & 105 & 273 & 0 & 0 & 0 \\ 0 & 0 & 0 & 112 & 0 & 0 \\ 0 & 0 & 0 & 0 & 118 & 0 \\ 0 & 0 & 0 & 0 & 0 & 98 \end{pmatrix}$$

density = 3.49 g/cm^3
symmetry: orthorhombic

γ-Spinel (γ - Mg_2SiO_4) (1)

$$(C_{ij}) = \begin{pmatrix} 327 & 112 & 112 & 0 & 0 & 0 \\ 112 & 327 & 112 & 0 & 0 & 0 \\ 112 & 112 & 327 & 0 & 0 & 0 \\ 0 & 0 & 0 & 126 & 0 & 0 \\ 0 & 0 & 0 & 0 & 126 & 0 \\ 0 & 0 & 0 & 0 & 0 & 126 \end{pmatrix}$$

density: 3.56 g/cm^3
symmetry: cubic

Ilmenite (2)

$$(C_{ij}) = \begin{pmatrix} 472 & 168 & 70 & -27 & 24 & 0 \\ 168 & 472 & 70 & 27 & -24 & 0 \\ 70 & 70 & 382 & 0 & 0 & 0 \\ -27 & 27 & 0 & 106 & 0 & 0 \\ 24 & -24 & 0 & 0 & 106 & 0 \\ 0 & 0 & 0 & 0 & 0 & 152 \end{pmatrix}$$

density = 3.82 g/cm^3
symmetry: trigonal

Perovskite (3)

$$(C_{ij}) = \begin{pmatrix} 515 & 117 & 117 & 0 & 0 & 0 \\ 117 & 525 & 139 & 0 & 0 & 0 \\ 117 & 139 & 435 & 0 & 0 & 0 \\ 0 & 0 & 0 & 179 & 0 & 0 \\ 0 & 0 & 0 & 0 & 202 & 0 \\ 0 & 0 & 0 & 0 & 0 & 175 \end{pmatrix}$$

density: 4.408 g/cm^3
symmetry: orthorhombic

Stishovite (4)

$$(C_{ij}) = \begin{pmatrix} 453 & 211 & 203 & 0 & 0 & 0 \\ 211 & 453 & 203 & 0 & 0 & 0 \\ 203 & 203 & 776 & 0 & 0 & 0 \\ 0 & 0 & 0 & 252 & 0 & 0 \\ 0 & 0 & 0 & 0 & 252 & 0 \\ 0 & 0 & 0 & 0 & 0 & 302 \end{pmatrix}$$

density = 4.29 g/cm^3
symmetry: tetragonal

(*) The tensor components are given in a direct set of Cartesian axes (x_1, x_2, x_3) oriented as follows:

- (x_1, x_2, x_3) ⇔([100], [010],[001]) for cubic, tetragonal and othorhombic
- (x_1, x_2, x_3) ⇔([100], perpendicular to (010), [001]) for trigonal crystals.

(1) Weidner et al. (1984); (2) Levien et al. (1979) ; (3) Yeganeh-Haeri et al. (1989); (4) Weidner et al. (1982).

With the exception of two pyroxenes (bronzite, diopside) and olivine, all the other minerals listed in Table 3-1 have coefficients of anisotropy of the shear wave velocities higher than the coefficients of the compressional velocities. The largest differences in the P- and S-wave anisotropy coefficients are observed for micas, feldspars and several other minerals.

Table 3-2 presents the principal high-density mineral phases which may contribute by their single crystal anisotropy to the seismic anisotropy of the deeper mantle. The phase transformation of alpha-olivine to the denser polymorph possessing the spinel structure was first suggested by Bernal in 1936. The transformation to the beta-phase under conditions of 400-km discontinuity (approximately at 13 GPa and 1400°C, Jeanloz and Thompson, 1983, see Fig. 3-2) results in a density increase of about 9% and an increase of average velocities from 8.81 km/s to 9.47 km/s for P waves and from 4.98 km/s to 5.74 km/s for S waves. While the velocity anisotropy for S waves does not virtually change after the transformation, the P-velocity anisotropy drops from 24% to 17% (Fig. 3-3), which is still high enough to contribute to a large-scale anisotropy of the mantle beneath the 400-km discontinuity. The next transformation of the beta-phase to the gamma-phase only leads to an insignificant change of density and P and S velocities, but the anisotropy drops to about 4% and 8%, respectively.

"Ilmenite" $MgSiO_3$ is a trigonal high-pressure form of enstatite, a stable phase at the base of the transition region and the top of the lower mantle (Anderson, 1989). This author assumes that due to its platy shape it may easily be oriented in the mantle. Due to this preferred orientation and the high single-crystal anisotropy (21% and 37% for P and S waves, respectively, see Table 3-2) a significant contribution of this mineral to the

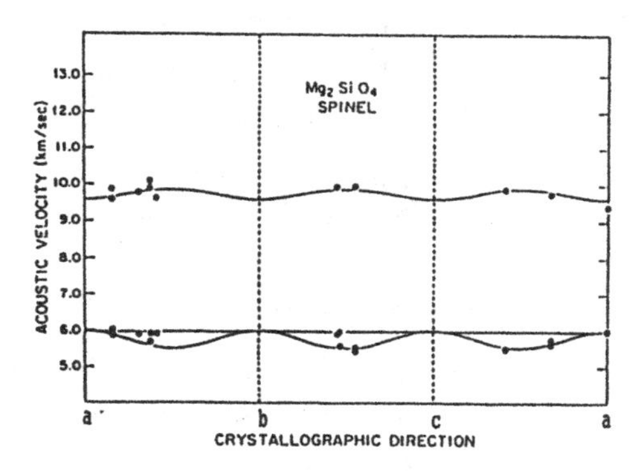

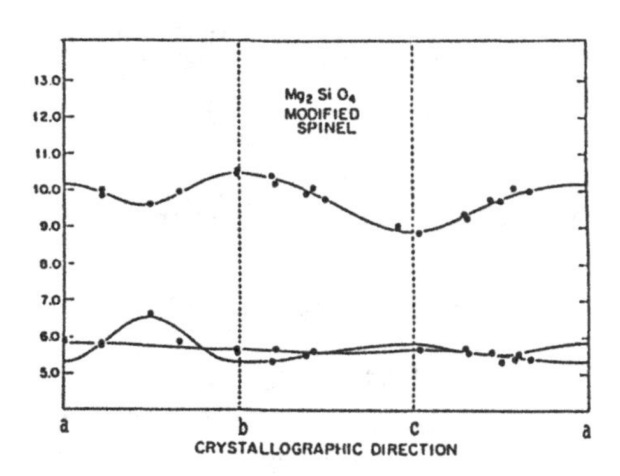

Figure 3-3 Acoustic velocities of spinel (gamma phase) and modified spinel (beta phase), Mg_2SiO_4 (Weidner et al., 1984). The solid lines are the calculated values based on the deduced elastic moduli. (AGU, copyright © 1984).

Table 3-3 Stiffness tensor components (*) of the major crustal and upper mantle minerals. Units are in GPa.

A - Major crustal minerals

Quartz (1)

$$(C_{ij}) = \begin{pmatrix} 86.6 & 6.7 & 12.6 & -17.8 & 0 & 0 \\ 6.7 & 86.6 & 12.6 & 17.8 & 0 & 0 \\ 12.6 & 12.6 & 106.6 & 0 & 0 & 0 \\ -17.8 & 17.8 & 0 & 57.8 & 0 & 0 \\ 0 & 0 & 0 & 0 & 57.8 & -17.8 \\ 0 & 0 & 0 & 0 & -17.8 & 39.9 \end{pmatrix}$$

density = 2.65 g/cm^3
symmetry: trigonal

Calcite (1)

$$(C_{ij}) = \begin{pmatrix} 144 & 53.9 & 51.1 & -20.5 & 0 & 0 \\ 53.9 & 144 & 51.1 & 20.5 & 0 & 0 \\ 51.1 & 51.1 & 84.0 & 0 & 0 & 0 \\ -20.5 & 20.5 & 0 & 33.5 & 0 & 0 \\ 0 & 0 & 0 & 0 & 33.5 & -20.5 \\ 0 & 0 & 0 & 0 & -20.5 & 45.1 \end{pmatrix}$$

density = 2.712 g/cm^3
symmetry:trigonal

Feldspar (microcline) (2)
$(Or_{78.5}Ab_{19.4}An_{2.1})$

$$(C_{ij}) = \begin{pmatrix} 62.5 & 42.8 & 3.8 & 0 & -15.4 & 0 \\ 42.8 & 172 & 24.1 & 0 & -14.3 & 0 \\ 3.8 & 24.1 & 124 & 0 & -11.5 & 0 \\ 0 & 0 & 0 & 14.3 & 0 & -2.8 \\ -15.4 & -14.3 & -11.5 & 0 & 22.3 & 0 \\ 0 & 0 & 0 & -2.8 & 0 & 37.4 \end{pmatrix}$$

density = 2.56 g/cm^3
symmetry: triclinic (quasi monoclinic)

Feldspar (oligoclase) (3)
(An_{24})

$$(C_{ij}) = \begin{pmatrix} 81.8 & 39.3 & 40.7 & 0 & -9.0 & 0 \\ 39.3 & 145 & 34.1 & 0 & -7.9 & 0 \\ 40.7 & 34.1 & 133 & 0 & -18.5 & 0 \\ 0 & 0 & 0 & 17.7 & 0 & -0.8 \\ -9.0 & -7.9 & -18.5 & 0 & 31.2 & 0 \\ 0 & 0 & 0 & -0.8 & 0 & 33.3 \end{pmatrix}$$

density = 2.64 g/cm^3
symmetry: triclinic (quasi monoclinic)

Diopside (4)

$$(C_{ij}) = \begin{pmatrix} 204 & 84.4 & 88.3 & 0 & -19.3 & 0 \\ 84.4 & 175 & 48.2 & 0 & -19.6 & 0 \\ 88.3 & 48.2 & 238 & 0 & -33.6 & 0 \\ 0 & 0 & 0 & 67.5 & 0 & 11.3 \\ 19.3 & 19.6 & 33.6 & 0 & 58.8 & 0 \\ 0 & 0 & 0 & 11.3 & 0 & 70.5 \end{pmatrix}$$

density = 3.31 g/cm^3
symmetry: monoclinic

Hornblende (5)

$$(C_{ij}) = \begin{pmatrix} 116 & 45 & 61 & 0 & 4 & 0 \\ 45 & 160 & 66 & 0 & -2 & 0 \\ 61 & 66 & 192 & 0 & 10 & 0 \\ 0 & 0 & 0 & 57.4 & 0 & -6 \\ 4 & -2 & -10 & 0 & 31.8 & 0 \\ 0 & 0 & 0 & -6 & 0 & 36.8 \end{pmatrix}$$

density = 3.124 g/cm^3
symmetry: monoclinic

Mica (muscovite) [6]

$$(C_{ij}) = \begin{pmatrix} 178 & 42.4 & 14.5 & 0 & 0 & 0 \\ 42.4 & 178 & 14.5 & 0 & 0 & 0 \\ 14.5 & 14.5 & 54.9 & 0 & 0 & 0 \\ 0 & 0 & 0 & 12.2 & 0 & 0 \\ 0 & 0 & 0 & 0 & 12.2 & 0 \\ 0 & 0 & 0 & 0 & 0 & 67.8 \end{pmatrix}$$

density = 2.790 g/cm^3
symmetry: monoclinic (quasi hexagonal)

B - Major upper-mantle minerals

Olivine [7]
$(Mg_{92.7}Fe_{7.2})_2SiO_4$

$$(C_{ij}) = \begin{pmatrix} 323.7 & 66.4 & 71.6 & 0 & 0 & 0 \\ 66.4 & 197.6 & 75.6 & 0 & 0 & 0 \\ 71.6 & 75.6 & 235.1 & 0 & 0 & 0 \\ 0 & 0 & 0 & 64.6 & 0 & 0 \\ 0 & 0 & 0 & 0 & 78.7 & 0 \\ 0 & 0 & 0 & 0 & 0 & 79.0 \end{pmatrix}$$

density = 3.311 g/cm^3
symmetry: orthorombic

Orthopyroxene [8]
(bronzite)

$$(C_{ij}) = \begin{pmatrix} 230 & 70 & 57.3 & 0 & 0 & 0 \\ 70 & 165 & 50 & 0 & 0 & 0 \\ 57.3 & 50 & 206 & 0 & 0 & 0 \\ 0 & 0 & 0 & 83.1 & 0 & 0 \\ 0 & 0 & 0 & 0 & 76.4 & 0 \\ 0 & 0 & 0 & 0 & 0 & 78.5 \end{pmatrix}$$

density = 3.335 g/cm^3
symmetry: orthorombic

Garnet (pyrope) [5]

$$(C_{ij}) = \begin{pmatrix} 287.4 & 105.0 & 105.0 & 0 & 0 & 0 \\ 105.0 & 287.4 & 105.0 & 0 & 0 & 0 \\ 105.0 & 105.0 & 287.4 & 0 & 0 & 0 \\ 0 & 0 & 0 & 91.6 & 0 & 0 \\ 0 & 0 & 0 & 0 & 91.6 & 0 \\ 0 & 0 & 0 & 0 & 0 & 91.6 \end{pmatrix}$$

density = 3.582 g/cm^3; symmetry: cubic

(*) The direct set of mutually perpendicular axes (x_1, x_2, x_3) is oriented as follows:

- symmetry around axis x_3 for hexagonal crystals
- (x_1, x_2, x_3) $\Leftrightarrow$([100], [010], perpendicular to (001)) for monoclinic crystals.

(for other symmetry systems see the caption of table 3-2).

(1) Hearmon (1956); (2) Ryzhova and Aleksandrov (1965); (3) Ryzhova (1964);
(4) Aleksandrov et al. (1963); (5) Hearmon (1979); (6) Aleksandrov and Ryzhova (1961c);
(7) Kumazawa and Anderson (1969); (8) Kumazawa (1969).

seismic anisotropy of the mantle might be expected. Anderson (1989) assumes that the "ilmenite" form of enstatite, a lower temperature phase than majorite (cubic high-pressure phase of pyroxene), appears mainly in deeply subducted slabs and contributes to the high velocities found in the vicinity of deep-focus earthquakes.

The high-pressure phase of SiO_2, stishovite, is characterized by extremely high velocities and anisotropy (Table 3-2 and Fig. 3-4). Due to its tetragonal symmetry and the orientation of the P-velocity extremes along [001] and [100] axes we may expect a very high anisotropy of stishovite aggregates which were strongly deformed at depth. Such situations can occur, e.g., in continent-continent collision zones like in the Alps, where stishovite was found by Chopin (1987). Stishovite might be also a form of silica in the Earth's transition zone and lower mantle, where, if systematically oriented, could contribute to velocity variations and to a large scale anisotropy (Anderson, 1987).

Another important high pressure mineral is perovskite $(Mg,Fe)SiO_3$, which together with magnetowustite (Mg,Fe)O is thought to be the most abundant phase in the lower mantle (Yeganeh-Haeri et al., 1989; Knittle et al., 1986). Perovskite is of an orthorhombic symmetry and exhibits elastic anisotropy (see table 3-2). P- and S-velocity anisotropies are 9.3% and 7.2%, respectively, as calculated from the data by Yeganeh-Haeri et al. (1989).

In order to model seismic-wave propagation in an arbitrary direction, we need the complete elastic tensor components. Elastic tensor components for the major minerals

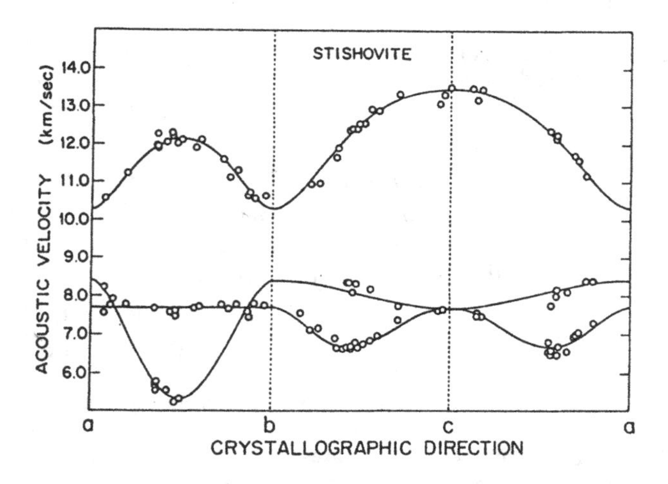

Figure 3-4 Acoustic velocities of stishovite (Weidner et al., 1982). The measured values are projected onto the nearest principal plane. The solid lines are the calculated values based on the deduced elastic moduli. (AGU, copyright © 1982).

of the crust and upper mantle are given in Table 3-3. Additional elastic constants as well as the temperature and pressure derivatives for some minerals are compiled in the handbooks by Simmons and Wang (1971), Carmichael (1984) and Landolt-Börnstein (1982). Elastic constants of several deep-mantle minerals can be found in Anderson (1989).

3.3 Elastic anisotropy of polycrystalline aggregates

Elastic properties of crystalline aggregates are treated in terms of the properties of the component crystals in respect of their arrangement, orientation and conditions along their contacts. Preferred mineral orientation, mineral layering and anisotropic cracks distribution are the main causes of the anisotropy of physical properties of rocks.

An initially isotropic solid can become anisotropic when acted on by sufficiently large stresses . However, the stresses required to cause anisotropic effects are probable only immediately adjacent to a potential earthquake (Evans, 1984a). In most cases the anisotropy induced by a non-hydrostatic stress field is connected with the opening or closing of systems of microcracks. While the cracks and faults play an important role in the anisotropy of the uppermost crust, the plastic-flow induced preferred orientation of minerals is believed to be a major cause of seismic anisotropy in deeper parts of the Earth. In the following section we will examine more closely these two dominant causes of the anisotropy.

3.3.1 Effects of cracks and grain boundaries

Practically all rocks, even dense crystalline varieties, contain small cracks. A number of laboratory investigations of rock samples have demonstrated that the crack porosity decreases the elastic wave velocities measured at atmospheric and low hydrostatic pressures (e.g. Birch, 1961). A spatial distribution of cracks exerts a strong influence on the elastic anisotropy of rocks. Oriented systems of cracks may cause a velocity anisotropy even if the matrix of the rock is almost isotropic (e.g., Nur and Simmons, 1969a; Babuska et al., 1977). The most prominent velocity reductions occur in the direction normal to the plane of oriented cracks.

The effects of the orientation of microcracks and grain boundaries on the anisotropy of P-wave velocities in samples of granodiorite and quartzite will be briefly demonstrated in this section. The elastic anisotropy was measured in spherical rock samples which enable the spatial distribution of velocities to be investigated, the dimensions and orientations of the cracks and grain boundaries were determined optically by universal stage techniques. Details of this investigation were presented in Babuska et al. (1977) and in Babuska and Pros (1984).

The anisotropy diagrams, illustrated by isolines of velocities in equal-area projection, are shown in Fig. 3-5a,b. The P-velocity anisotropy of the granodiorite (Sazava river, Czechoslovakia) at atmospheric pressure is remarkable - the maximum difference in velocities is almost 1.6 km/s. After the application of hydrostatic pressure, both the velocities and the anisotropy changed significantly. Already at a pressure of 20 MPa, which corresponds to a depth of about 800 m, the maximum difference in velocities is only 0.66 km/s and, in the direction of the velocity minimum, the velocity increased by as much as 1.6 km/s. The increase of velocities and the decrease of anisotropy also continued at higher pressures up to 300 MPa (equivalent to a depth of about 11 km, Fig. 3-5b) when the maximum difference between extreme velocities is 0.27 km/s. The highest increment of velocity was observed between atmospheric pressure and 20 MPa. The velocity pattern under atmospheric conditions, after the removal of the pressure, is very similar to the pattern at the beginning of the measurement; however, the maximum velocity is about 180 m/s and the minimum velocity about 300 m/s higher than at the beginning of the measurement. This hysteresis indicates irreversible changes in the rock structure, namely that a certain number of microcracks remain closed.

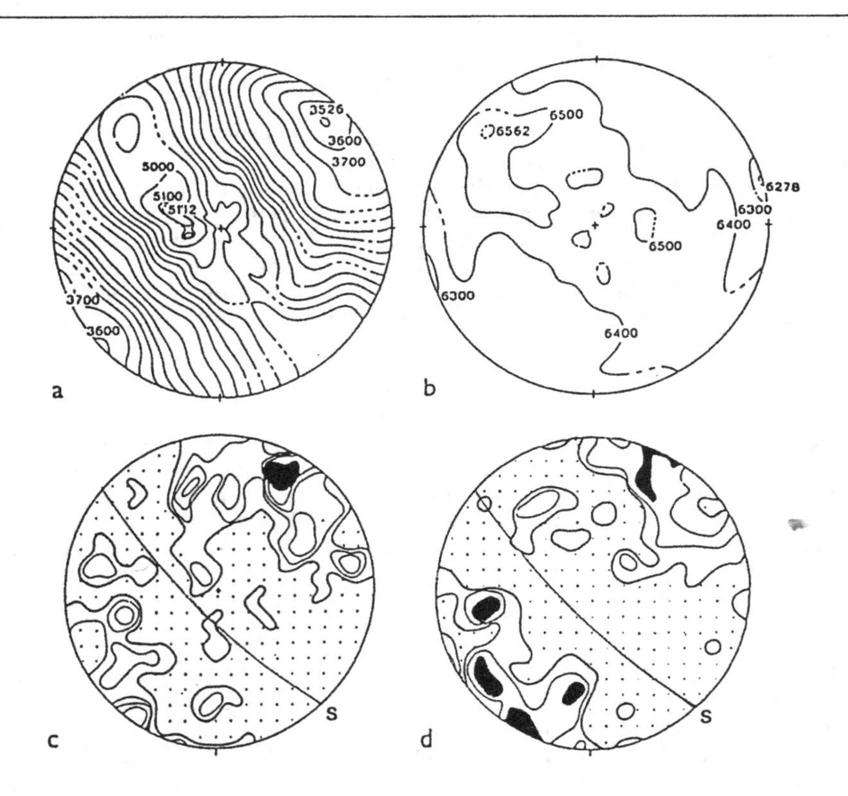

Figure 3-5 River Sazava granodiorite (Babuska and Pros, 1984): diagrams of elastic anisotropy (a,b) and orientation diagrams of cracks in hornblende (c) and biotite (d). (a) Initial measurement at atmospheric pressure, (b) hydrostatic pressure of 300 MPa. In both diagrams isolines of compressional velocities are constructed in steps of 100 m/s and extreme values are given. (c) Orientation of perpendiculars to (110) cleavage in 72 hornblende grains (contours 5,4,3,2 and 1% per 1% area). (d) Orientation of perpendiculars to (001) cleavage in 143 biotite grains (contours 3, 2 and 1% per 1% area). S indicates the foliation plane. (With permission of the Royal Astronomical Society, copyright © 1984).

These velocity patterns demonstrate that, while the investigated granodiorite is strongly anisotropic at atmospheric pressure, it is nearly isotropic at 300 MPa. This observation suggests that most of the observed anisotropy is caused by the anisotropic distribution of cracks in the rock. To compare the elastic anisotropy of the granodiorite with the rock fabric, we measured the preferred orientation of four minerals which form about 84% of the rock volume, the orientation of the c-axes in quartz grains and the cleavage cracks in plagioclase, hornblende and biotite. No direct relation between the preferred orientation of quartz and the elastic anisotropy of granodiorite was observed. Cracks in quartz grains are infrequent and their orientation seems to be distributed randomly. They may contribute to the reduction of the velocities at atmospheric pressure, but they do not have any substantial effect on the anisotropy.

On the other hand, the perpendiculars to the cleavage cracks of biotite (Fig. 3-5c) and hornblende (Fig. 3-5d) agree with the direction of minimum velocities at atmospheric and low hydrostatic pressures. At higher pressures, when these cracks are mostly closed, their influence on the P-wave velocities is reduced, and the anisotropy is due to the lattice and shape orientations of the minerals. Biotite and hornblende each form about 9% of the rock volume. It is, therefore, understandable that, in spite of a well-defined preferred orientation of these two highly anisotropic minerals (Fig. 3-5c,d), the velocity difference at a pressure of 300 MPa is so small (less than 0.3 km/s). At atmospheric pressure, the cleavage cracks of biotite and hornblende are opened enough to reduce the P velocities in the direction approximately perpendicular to the majority of the cleavage planes from 6.28 km/s at 300 MPa to only 3.53 km/s. As was discussed in Babuska et al. (1977), the cleavage system in plagioclase does not play any important role in the anisotropy of the investigated granodiorite.

The other investigated rock, quartzite (Loket, Czechoslovakia), is a massive, almost monomineral material with an admixture of 3% of fine-grained biotite and chloritized pyroxene. The orientation of 440 microcracks and 315 grain boundaries in three mutually perpendicular thin sections was determined (Babuska and Pros, 1984). In comparison with granodiorite, the quartzite sample shows a lower anisotropy and a symmetry of the isoline pattern at atmospheric pressure and also a smaller change of velocities when hydrostatic pressure is applied. Fig. 3-6a,b shows how the compressional wave velocities change as the pressure increases from the atmospheric to 400 MPa. A general pattern of velocities in space is preserved in the whole pressure range. The maximum velocity increased from 5.62 to 5.89 km/s.

Figure 3-6d demonstrates the difference in velocities determined for the individual directions at the same pressure of 0.1 MPa for the increasing and the decreasing pressure cycles. The diagram thus displays elastic hysteresis which attains its maximum value of 120 m/s in the direction close to the statistically determined maximum of the perpendiculars to the grain boundaries (Fig. 3-6e) and microcracks (Fig. 3-6f). The configuration of isolines for both types of microdiscontinuities in the rock is similar. A great degree of conformity is also observed for the space orientation of cracks of different dimensions. However, the smallest cracks show a more scattered distribution.

The results of the quantitative correlation (Babuska and Pros, 1984) demonstrate that the nature of the microdiscontinuities in rocks plays a very important role in their effects on the P velocities, namely at low confining pressures. Well-developed cleavage cracks in biotite and amphibole have a greater influence on the velocity anisotropy than

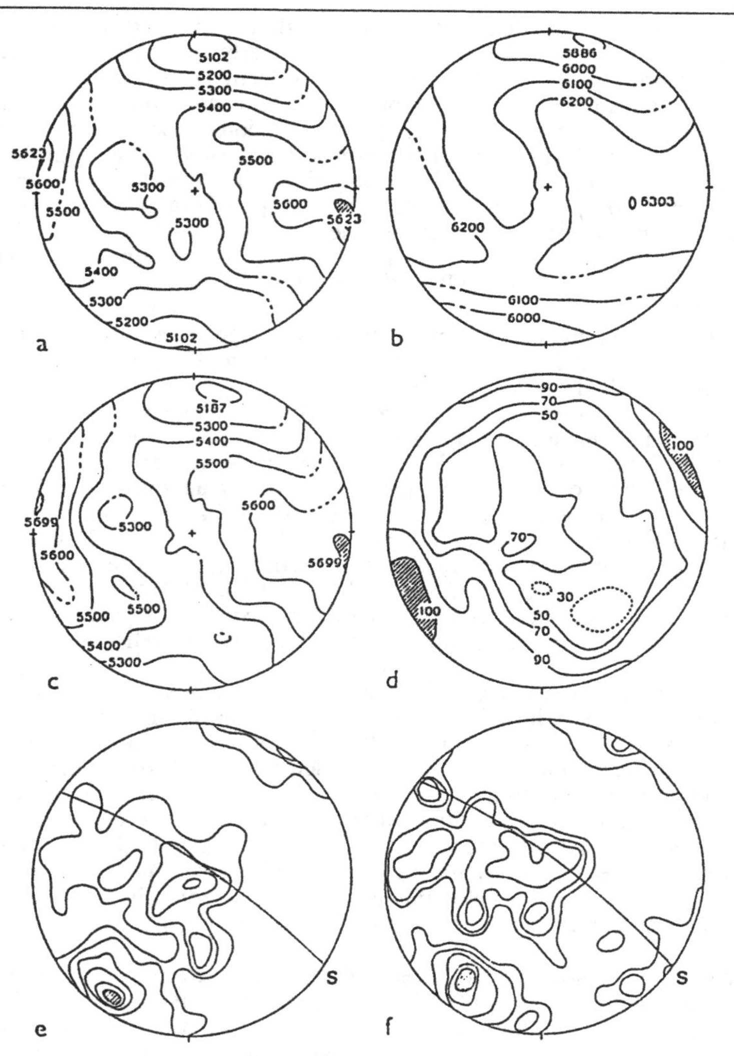

Figure 3-6 Quartzite from locality Loket, Czechoslovakia (Babuska and Pros, 1984). Diagrams of elastic anisotropy demonstrate changes of compressional velocities from: (a) atmospheric pressure to (b) the maximum applied hydrostatic pressure of 400 MPa, and (c) the atmospheric pressure at the end of the cycle. (d) Difference between the velocities at atmospheric pressure after the experiment and the velocities in the virgin state (hysteresis). The bottom diagrams show statistical orientations of normals to the grain boundaries (e) and microcracks (f). S is the foliation plane. (With permission of the Royal Astronomical Society, copyright © 1984).

numerous but less pronounced cleavage cracks in plagioclase. In comparison with granodiorite, quartzite showed a less ordered distribution of microcracks. The microdiscontinuities in quartzite are closed at a pressure of about 100 MPa, unlike granodiorite, in which the closing of cracks would probably continue at pressures over 300 MPa, which corresponds to a depth of about 11 km.

Shear-wave velocities also distinctly decrease for propagation normal to the plane of preferred orientation of cracks, but relatively less than compressional velocities (independent of polarization direction), and are affected to the same degree for propagation in the preferred plane if the polarization is normal to this plane. For propagation and polarization in the preferred plane of the cracks the influence on the shear velocity is very small (Gebrande, 1982).

In natural conditions open cracks are usually filled with fluids (e.g., Brace et al., 1966). P velocities strongly increase and P-wave anisotropy decreases when cracks are water saturated. S-wave velocities and anisotropy are almost independent on water saturation. Under dry conditions the anisotropy due to cracks is greater for P waves than for S waves whereas this dependence is reversed for water-saturated cracks (Paterson, 1978). A pressure of pore fluids can keep the cracks open at any depth and thus decrease seismic velocities. Kern and Schenk (1985) observed that in rocks which represent a lower crustal section in southern Calabria, most of the microcracks are closed at a confining pressure of about 200 MPa. At higher pressures, preferred lattice orientation is the dominant cause of the observed velocity anisotropy.

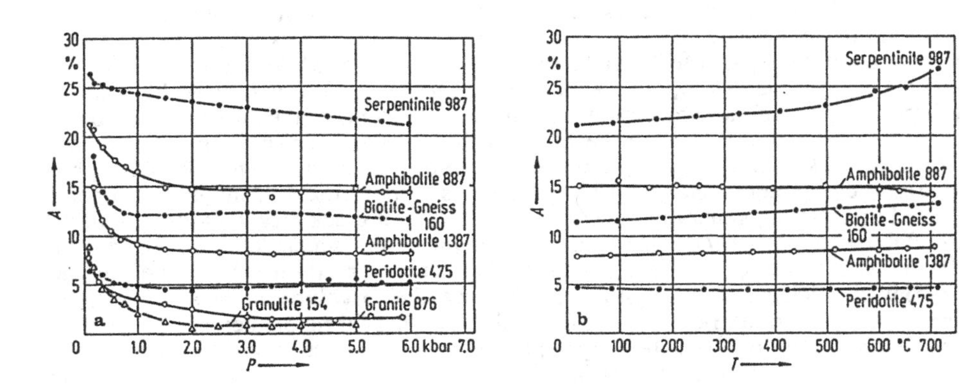

Figure 3-7 P-velocity anisotropy (A) in several types of crustal and mantle rocks as a function of pressure (P) at room temperature (a) and as a function of temperature (T) at 600 MPa confining pressure (b) (Kern, 1982). (With permission of Springer International, copyright © 1982).

While the pressure effects on the anisotropy of rocks caused by the distribution of cracks is very important, in general, the velocity anisotropy is only insignificantly affected by temperature at a high confining pressure (Fig. 3-7). Kern and Schenk (1988) observed that at the pressure of 600 MPa and temperature of 600°C a significant anisotropy, caused by the preferred orientation of minerals, remains for P and S velocities. Maximum values are 10.5% and 11.5% in ductile mylonites, 10.7% and 11.5% in gneisses, 15% and 13% in amphibolites.

The behavior of a serpentine-bearing peridotite was an exception (Kern and Schenk, 1985). In that rock the P-velocity anisotropy increased from about 11% up to about 20% at 550°C as a result of a breakdown of the serpentine. The decrease of P-wave velocities normal to the foliation plane may be explained by the breakdown of the weaker bond between the layers of this sheet silicate whereas the stronger bonds within the layers are not immediately affected (Kern and Fakhimi, 1975).

Oriented cracks are important in seismological and tectonic investigations of the upper crust as they may indicate the effect of prevailing stress on pre-existing fractures or microcracks (see section 5.4.1).

3.3.2 Effects of preferred mineral orientation

At a high confining pressure, when the effect of microcracks on seismic velocities is eliminated, velocity anisotropy is entirely controlled by the preferred orientation of major minerals, which can be demonstrated by the universal-stage measurements of mineral orientations.

Let us take from Table 3-1 the most abundant minerals or mineral groups of crystalline rocks and arrange them in a series according to the values of the coefficients of the compressional-wave anisotropy k. This results in the following sequence: micas (k=59%), alkali feldspars (46%), calcite (33%), plagioclase (31%), hornblende (27%), quartz (26%), olivine (25%), pyroxenes (16%), garnets (1%). It is remarkable that the last three members of the sequence are the main components of the upper mantle rocks. The olivine rich rocks like dunite and peridotite, however, belong to the most anisotropic rocks. In contrast, granite or granodiorite, being composed of highly anisotropic components and often carrying signs of deformation, including mineral preferred orientation, generally have low coefficients of crack-free anisotropy (Birch, 1961; Babuska et al., 1977). What then are the main factors which determine the degree of the elastic anisotropy of crystalline rocks? There are three: the values of the anisotropy coefficients of the constituent minerals and their volume in the aggregate; the degree of the preferred orientation of minerals and other fabric elements, e.g. oriented microdiscontinuities or alternating layers; the orientation of active slip directions - like [100] in olivine (Gueguen and Darot, 1982) - with respect to the symmetry of the single-crystal elastic anisotropy. Besides garnets, whose anisotropy is negligible, all other very abundant minerals named above can contribute, according to their single crystal anisotropy, to the elastic anisotropy of rocks.

The main mechanism systematically orienting minerals and causing the velocity anisotropy of their aggregates is plastic and viscous flow (Mainprice and Nicolas, 1989). In a plastic field, crystals deform principally by the dislocation slip. Simple slip systems occur in olivine (Gueguen and Nicolas, 1980) and allow significant preferred orientations to be developed. In this chapter we selected only the principal rock-forming minerals and their orientations: for details the reader is referred to specialized literature.

Deformation experiments on the preferred orientation of mica carried out by Tullis (1971) showed that during crystallization or recrystallization the [001] plane mean orientation was always normal to the direction of the greatest shortening which, in experiments, coincides with the maximum compressive stress σ_1. The compressional velocity minimum in micas is perpendicular to the [001], which is also an important slip system plane and the velocity maximum is parallel to this plane (see Table 3-1). Due to these orientations and to the extremely high crystal anisotropy, micas contribute substantially to the elastic anisotropy of aggregates. This is true of a number of metamorphic rocks such as mica schists and gneisses.

Minerals of the feldspar group constitute over 51% of the crust volume (Ronov and Yaroshevsky, 1969). The lattice preferred orientation studies of plagioclase have received little attention though it is a major mineral of the upper and lower crust, particularly in high-grade metamorphic rocks. Therefore, it has been stated that, to a first approximation, the rheology of the crust could be taken to be the rheology of feldspar (Tullis, 1979). Ji and Mainprice (1988) studied microstructures and fabrics of twelve naturally deformed plagioclase-containing rocks and concluded that (010) as the dominant slip plane tended to be parallel to the foliation and that the asymmetry of the (010) pole figure with respect to foliation and lineation indicated that the rock was deforming in a non-coaxial regime and could be used to determine the sense of shear. Either [100] or [001] form concentrations sub-parallel to the lineation of rocks (Mainprice and Nicolas, 1989). For such a typical orientation we would expect the plagioclase anisotropy to be dominant in the plagioclase-rich rocks of the lower crust. However, Ji and Mainprice (1988) pointed out that other orienting mechanisms, like the superplastic deformation, were important in the lower crust. One of the characteristics of this mechanism is that it does not produce a preferred orientation and it can randomize a pre-existing fabric. Hence, no seismic anisotropy is expected in plagioclase rocks deformed by this mechanism and the variation between regions of strong mineral orientations and superplastic shear zones should produce excellent seismic reflectors (Ji and Mainprice, 1988).

Experimental deformation of calcite aggregates was investigated in a number of studies which can be summarized as follows: calcite deforms by e-twinning at low temperatures which rotates the c-axes into near parallelism with σ_1. This regime was observed for Carrara marble deformed at a temperature of 700°C and for Solnhofen limestone at room temperature (Schmid et al., 1987). At higher temperatures glide planes are activated and the c-axis orientations result from an alignment of slip planes and slip directions. Since the c-axis is the direction of the velocity minimum (also the direction of high compressibility, Birch, 1966) and the P-wave velocities in the x - y plane are very close to the maximum velocity, the directions of the velocity extremes are often preferentially oriented, which leads to the commonly observed high elastic anisotropy of marbles.

The preferred orientation of coarse-grained hornblende in metamorphic rocks is often observed. Schwerdtner (1964) found that the c-axes of hornblende in a deformed gneiss from Norway tended to be significantly oriented. The c-axis is also the direction of maximum P-wave velocity and thus the oriented hornblende grains can contribute to an increase of the elastic anisotropy of amphibolites in the lower crust (Siegesmund et al., 1989; Luschen et al., 1990). Both in hornblende and in micas the external shape plays an important role in orientation and the resulting anisotropy of metamorphic complexes (Burdock, 1980), because after the rotation the velocity maxima of both minerals are oriented along the foliation.

The investigation of the preferred orientation of quartz shows a complex picture of the different types of orientation developed under different experimental conditions. Dell'Angelo and Tullis (1989) deformed a fine-grained quartzite to varying amounts of shear strain. They studied the progressive development of deformation microstructures, including planar fabrics and c-axis preferred orientation, as a function of increasing shear strain. The c-axis preferred orientations of original grains define symmetric crossed circle girdles at low strain and asymmetric broad maxima at high strain, with the greater concentration of c-axes in the direction of shear.

Taking into consideration the complexity of the slip systems and various quartz orientations, as well as the orientations of the velocity extremes in quartz (Table 3-1), we cannot expect a very pronounced elastic anisotropy of rocks due to the orientation of quartz grains. Mainprice and Nicolas (1989) determined the P-velocity anisotropy of 6% for a quartzite computed from the quartz fabric derived from the neutron diffraction analysis. Quartz mylonites produced by intense ductile deformation in shear zones which may act as seismic reflectors in the lower continental crust, can attain P-velocity anisotropies between 8 and 11% (Mainprice and Casey, 1990).

It is now generally accepted that seismic anisotropy in the mantle is due primarily to the deformation-induced lattice preferred orientation of olivine crystals. Among the mechanisms producing the preferred orientation under the upper-mantle conditions the dislocation creep seems to play the most important role, at least at the high strains associated with the flow near ridges and subduction zones (Ribe, 1989). Karato et al. (1980) reviewed the role of dynamic recrystallization and conclude that it may be very important, together with intracrystalline slip, in generating preferred orientations in the mantle.

Olivine in plastically deformed aggregates tends to be oriented with the [010] direction parallel to the flow plane and with [100] parallel to the flow line. This orientation confirms the experiments on plastic flow or shear along transverse faults, which were discussed e.g. by Nicolas and Poirier (1976) and which lead to a distinct preferred orientation of olivine crystals. At very high temperatures, the preferred orientation of olivine can be extremely strong due to enhanced diffusion and grain boundary mobility selecting the favorably oriented grains (Mainprice and Nicolas, 1989). Olivine has a P-velocity anisotropy of about 25% with the maximum velocity along the [100] direction and the minimum velocity along [010]. Such favorable conditions and the existence of a simple slip system in olivine result in the fact that a significant elastic anisotropy of olivine-rich ultramafic rocks is a rule rather than an exception.

Orthopyroxene is common in lherzolites and harzburgites.The experiments of Carter et al. (1972) on syntectonic recrystallization of orthopyroxene (enstatite) and natural deformation fabrics showed that in orthopyroxene [100] is parallel to the flow plane and [001] to the flow line. The fabric is similar to that of olivine. The orientation of velocity extremes in orthopyroxene (bronzite, see Table 3-1) is conformable to olivine. According to Kumazawa et al. (1971), provided olivine and orthopyroxene recrystallized in the same stress field, the anisotropies due to both minerals are additive for the anisotropy of rocks in bulk. However, Helmstaedt et al. (1972) observed in a nodule of lherzolite, and Peselnick et al. (1974) in a mantle peridotite from the Ivrea zone, that the crystallographic a-axes of enstatite and those of olivine form maxima perpendicular to each other, corresponding to the contrasting non-additive anisotropies of both minerals. Also Christensen and Salisbury (1979) confirmed that in ultramafic rocks the P-velocity maximum of orthopyroxene parallels the P-velocity minimum of olivine and thus dilutes the overall anisotropy of olivine-rich rocks.

As orthopyroxene is less ductile than olivine (Gueguen and Nicolas, 1980), the preferred orientations of orthopyroxene are often weaker than those of olivine (Nicolas and Poirier, 1976) and the anisotropy coefficients of orthopyroxene are smaller. The resulting anisotropy of orthopyroxene-rich rocks is relatively low (Babuska, 1972b).

Kumazawa et al. (1971) observed that, despite a significant preferred orientation of clinopyroxene, the velocity anisotropy of eclogites from diatremes of the Colorado plateau is considerably smaller than the anisotropy of peridotites. Although the coefficients of anisotropy of the omphacitic pyroxene have not been determined yet, we can assume a value of about 20% as for augite. On the basis of the thermodynamic theory of Kumazawa (1963), Kumazawa et al. (1971) assume that the preferred orientation of clinopyroxenes is controlled by their linear compressibility. Since in clinopyroxenes the directions of extreme linear compressibilities are different from the directions of the compressional velocity extremes, a high velocity anisotropy in eclogites cannot be expected. Such a dependence of orienting mechanisms on linear compressibilities has not been confirmed by experiments. However, complicated systems of clinopyroxene deformation and recrystallization (Gueguen and Nicolas, 1980), in combination with the low crystal symmetry, seem to be unfavorable for creating a high anisotropy of clinopyroxene aggregates.

Although the shear-wave anisotropy of most rock-forming minerals is larger than the compressional anisotropy, the shear anisotropy of aggregates is a complicated problem because of the great differences in the shear velocities with different polarizations. For example, in olivine the maximum difference due to polarization for a single direction of propagation is as much as 0.46 km/s (Kumazawa and Anderson, 1969) and in dunite, which is composed of 98% of olivine, the difference for a single direction amount to 0.38 km/s (Babuska, 1972b), this means 10% and 8%, respectively, of the mean velocity in the corresponding directions. The shear-wave birefringence was also observed in a number of metamorphic rocks (Tilmann and Bennet, 1973) and polarization studies using three-component records show diagnostic anomalies in the presence of in situ anisotropy (Keith and Crampin, 1977).

3.4 Crustal rocks

Most of the rocks constituting the crust are anisotropic due to preferred orientations of minerals, small-scale compositional layering and, in the upper crust, also due to the oriented systems of microcracks.

In general, most sediments can be described as transversely isotropic media. While the velocities vary insignificantly in dependence on direction in massive rocks like sandstones, many sediments with a plane parallel structure exhibit substantial anisotropy

Table 3-4 Density, seismic velocities and seismic anisotropy k of typical crustal rocks

Rock type	Number of samples	Range of density [g/cm^3]	Range of mean vp [km/s]	Average of mean vp [km/s]	Range of anisotropy [k%]	Average anisotropy [k%]	Hydrost. pressure [GPa]	Source of data
1 Granite	14	2.62-2.88	5.81-6.61	6.40	0.3-3.8	1.6	1	Birch,1960a Bayuk et al., 1982
2 Quartz diorite	3	2.76-2.91	6.60-6.71	6.67	0.1-1.1	0.6	1	Birch, 1960a
3 Gabbro (norite, anorthosite)	13	2.77-3.05	6.91-7.59	7.28	0.7-6.2	2.8	1	Christensen, 1978 Birch, 1960a Kroenke et al., 1976, Bayuk et al., 1982
4 Gabbro (norite, labradorite)	5	2.96-3.21	6.68-7.48	7.06	0.7-4.5	2.6	0.5	Bayuk, 1966 Bayuk et al., 1982
5 Basalt	4	2.79-2.83	6.09-6.54	6.36	0.3-2.2	1.2	1	Bayuk et al., 1982
6 Diabase	5	2.97-3.01	6.63-6.93	6.84	0.1-1.9	0.6	1	Birch, 1960a,
7 Schist	18	2.68-3.10	5.83-7.48	6.88	0.8-17.7	10.0	1	Bayuk et al., 1982 Galdin, 1977
8 Schist	8	2.75-3.19	6.59-7.54	6.96	1.1-21.0	10.2	1	Birch, 1960a Christensen, 1965
9 Gneiss	25	2.64-3.16	5.73-8.54	6.60	2.2-21.6	6.7	1	Bayuk et al., 1982 Galdin, 1977
10 Gneiss	9	2.64-2.85	6.30-6.58	6.43	0.3-6.8	3.5	1	Birch, 1960a Christensen, 1965
11 Amphibolite	8	2.96-2.29	6.35-8.27	7.13	0.7-19.8	9.5	1	Levykin, 1974 Bayuk, 1966 Bayuk, et al., 1971 Galdin,1977 Volarovich et al., 1977
12 Amphibolite	6	3.03-3.26	6.85-7.75	7.29	1.7-14.2	8.7	0.6-1	Birch, 1960a Christensen 1965 Siegesmund et al., 1989
13 Granulite	27	2.67-3.74	6.48-7.48	6.96	0.6-5.6	2.4	1	Christensen and Fountain 1975 Manghnani et al., 1974
14 Serpentinite	4	2.60-2.80	6.00-6.84	6.49	0.8-11.3	4.6	1	Birch, 1960a

with greater velocities parallel to bedding. Anisotropy may range up to 30% and is particularly pronounced for P-wave velocities from about 2.0 to 4.2 km/s (Gebrande, 1982).

Velocities and their anisotropy depend, namely in sediments, on a degree of saturation of the pore space with fluids and on the pressure of the pore fluid. Water saturation increases the P velocities at small pressures and all the more the greater the crack porosity of the rock is (Nur and Simmons, 1969b). The difference decreases with increasing pressure and may be neglected above about 0.2 GPa which corresponds to a depth around 7-8 km. In contrast to P-wave velocities, S-wave velocities are barely influenced by the pressure of pore fluids with zero pore pressure. If, however, the pore pressure equals the confining pressure, microcracks and grain boundaries are kept opened and P- as well as S-wave velocities remain much smaller with increasing pressure than the values measured in dry rocks.

Glacier ice covers about 10% of the Earth's surface at present, therefore, it has to be regarded as a common naturally occurring crustal rock. Since a single crystal of ice is slightly anisotropic (Miller, 1982) and the preferred orientation of c-axes is often developed (Lliboutry and Duval, 1985), we can expect velocity anisotropy in many glaciers and polar ice caps. An anisotropic travel time analysis has been used to measure the width of ice fabrics in Antarctic ice caps (Blankenship et al., 1989).

Table 3-4 reviews the published data on densities, compressional wave velocities and the crack-free anisotropy of the main groups of igneous and metamorphic crustal rocks. The average anisotropies of the individual groups of igneous rocks are very low at elevated hydrostatic pressures and only exceptionally do the anisotropy coefficients amount to several percent in some gabbros (Christensen, 1978). On the other hand, a number of metamorphic rocks are characterized by high anisotropy coefficients: schists and amphibolites with distinct foliations have coefficients k of about 20%. However, the average anisotropy of these two groups is between 8 and 10%. The extreme variability of the velocities and the anisotropy of gneisses are explained by the enormous differences in their compositions (Bayuk et al., 1982). It is remarkable that granitoids composed of highly anisotropic components, and often carrying signs of plastic deformation including mineral preferred orientations (Babuska et al., 1977), generally have low coefficients of crack-free anisotropy. As discussed in the previous section, this is caused by the directions of the velocity extremes in alkali feldspars, plagioclases and quartz differing from the directions which are decisive for their orientation during plastic flow. Thus, the velocity extremes are not aligned according to the directions of flow as in aggregates of micas and hornblende.

The velocity anisotropy is strongly dependent on rock fabrics. Kern and Wenk (1990) showed that anisotropy increased significantly with the degree of deformation whereas average velocities and densities did not change. At 600 MPa most of the intracrystalline cracks are closed and the residual anisotropy ranges for both P and S velocities from about 1% in granodiorite to about 12% in strongly deformed phyllonite (Fig. 3-8). During phyllonitization of granodiorite the fabric changes, namely the size, shape and orientation of mineral grains. The most significant is the preferred orientation of biotite.

As shown by Simmons (1964) and Christensen (1966a) the shear-wave anisotropy is very much dependent on rock fabrics as well. Experimental data of Kern

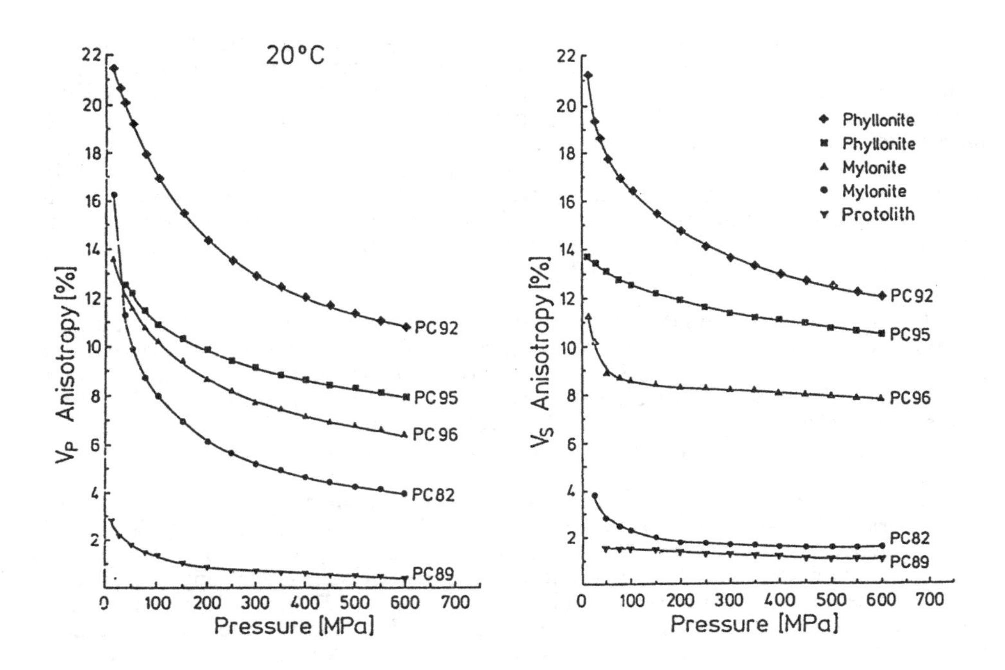

Figure 3-8 Velocity anisotropies in several rocks with very close average velocities and densities, but different degrees of deformation (Kern and Wenk, 1990). (AGU, copyright © 1990).

and Wenk (1990) in Fig. 3-9 show that significant shear-wave splitting is observed for the propagation parallel to the foliation plane. In the direction normal to the foliation plane the shear-wave splitting anisotropy is generally low. The decrease of this anisotropy correlates with the increase of velocities in the two perpendicularly polarized shear waves and is again attributed to a progressive closure of open cracks under increasing confining pressure.

Fig. 3-10 shows separately the ranges of the laboratory determined P-wave velocities which are given, first, by changes in composition of rocks corresponding to an average velocity, and secondly, by the anisotropy. Although the variations of velocities due to the anisotropy of schists, amphibolites and some gneisses are large, the variations due to changes in composition are substantially greater for all crustal rocks, especially igneous rocks. It is, therefore, also probable that the large-scale variability of P-wave velocities in the crust is mainly caused by inhomogeneities of different scales though the rock anisotropy plays locally an important role as well.

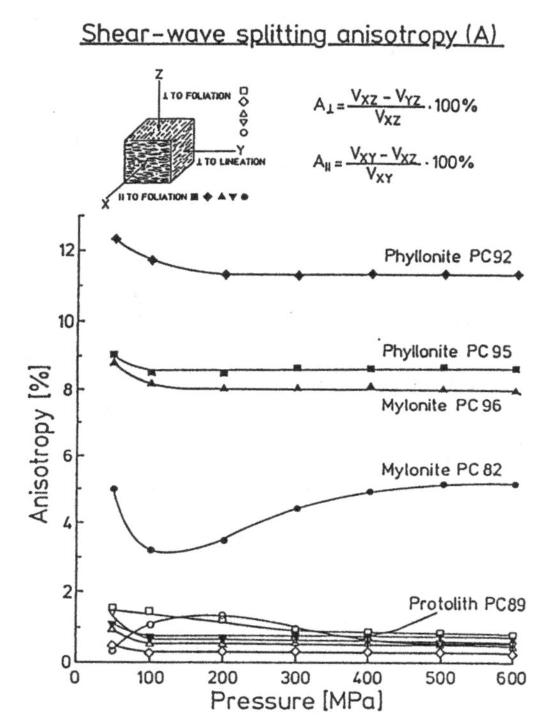

Figure 3-9 Measured shear-wave anisotropies as a function of the confining pressure in rocks with different degrees of deformation (Kern and Wenk, 1990). (AGU, copyright © 1990).

As the anisotropy of granulites is surprisingly small, amphibolites and gabbros are the main candidates for the anisotropic rocks in the lower crust. Anisotropic stress fields or a large-scale layering of inhomogeneities such as systems of dykes or fault zones can have an additional influence on the anisotropy of the crust.

3.5 Uppermost mantle rocks

As to the physical properties, namely densities and elastic wave velocities, peridotites, pyroxenites and eclogites are the most important representatives of the uppermost mantle materials. It is generally believed that olivine-rich ultramafic rocks are the main rock types of the upper mantle. Eclogites play an important role in subduction zones, where basaltic materials are partly returned to the Earth's interior. Schulze (1989) examined heavy mineral concentrates from eclogite-dominated kimberlites and

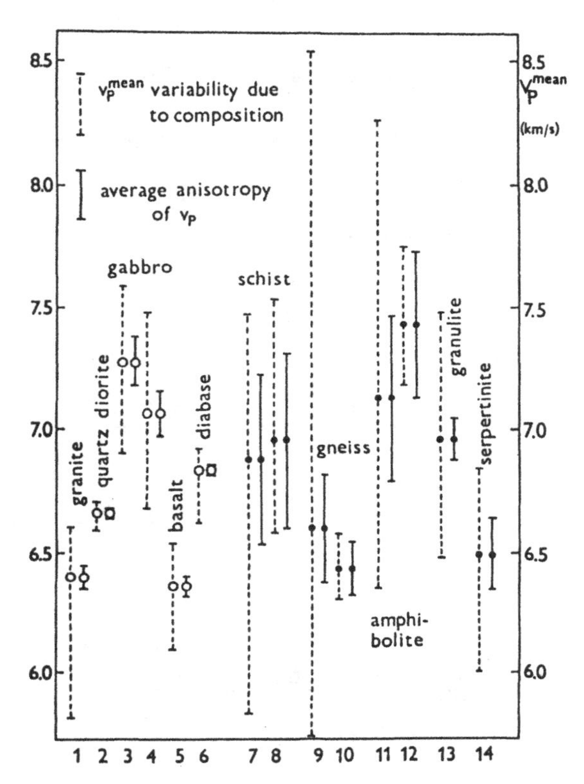

Figure 3-10 Mean velocities of compressional waves determined under high hydrostatic pressure in crustal igneous (open circles) and metamorphic (solid circles) rocks, and their variability due to compositional changes and anisotropy (Babuska, 1984). Numbers on the horizontal axis correspond to groups of samples in Table 3-3. (With permission of the Royal Astronomical Society, copyright © 1984).

concluded that the amount of eclogite in the upper 200 km of the subcontinental uppermantle might be even less than 1% by volume overall.

Peridotites are composed of olivine as the dominant mineral (from 40 to over 95%), two aluminous pyroxenes - enstatite and diopside, and chromiferous spinel in a minor quantity. In dependence on their mineral content, special varieties of peridotites are distinguished (dunite, olivinite, lherzolite, wehrlite, harzburgite). For pyroxenites, composed mainly of pyroxenes, special terms are used according to the dominant minerals : clinopyroxenites (diopsidite, diallagite), orthopyroxenites (enstatitite, bronzitite, hypersthenite) and websterites (both the clino- and orthopyroxenes form a substantial volume of the rock). Eclogites are deeply metamorphosed rocks of a basaltic

Table 3-5 Density, seismic velocity and seismic anisotropy k of upper mantle rocks.

Rock type	Number of samples	Range of density [g/cm^3]	Range of mean v_p [km/s]	Average of mean v_p [km/s]	Range of anisotropy [k%]	Average anisotropy [k%]	Hydrost. pressure [GPa]	Source of data
1 Dunite	8	3.26-3.33	8.00-8.66	8.32	7.2-15.0	10.8	1	Birch, 1960a Christensen, 1966b Babuska, 1972b
2 Lherzolite (xenoliths)	6	3.31-3.34	7.82-8.37	8.06	3.8-11.1	7.5	1	Babuska, 1976
3 Peridotite (xenoliths)	2	3.29	8.21-8.36	8.28	3.3-3.9	3.6	1	Christensen, 1966b
4 Peridotite (xenoliths)	2	3.23-3.32	7.92-8.08	8.00	11.2-12.5	11.8	0.5	Kumazawa et al. (1971)
5 Peridotite	2	3.28-3.29	7.70-8.74	8.29	5.0-9.1	7.0	1.5	Levitova, 1980
6 Olivinite	4	3.20-3.34	7.45-8.60	7.95	3.1-13.2	10.2	1.5	Levitova, 1980
7 Harzburgite	3	3.20-3.27	7.58-7.71	7.64	9.2-12.3	10.8	1	Kroenke et al., 1976
8 Pyroxenite	3	3.24-3.36	8.01-8.29	8.18	0.5-5.4	3.5	1	Birch, 1960a Kroenke et al., 1976
9 Bronzitite	3	3.21-3.30	7.83-8.02	7.91	1.3-6.2	3.9	1	Birch, 1960a Babuska, 1972b
10 Eclogite	11	3.23-3.64	7.90-8.61	8.24	0.7-3.9	2.3	1	Manghnani et al., 1974
11 Eclogite	3	3.34-3.45	7.69-8.02	7.86	1.2-2.9	2.2	1	Birch, 1960a
12 Eclogite	2	3.26-3.38	7.66-7.85	7.75	2.5-6.3	4.4	1	Galdin, 1977
13 Eclogite	9	3.19-3.53	7.55-8.39	8.01	0.7-3.8	2.0	0.75	Babuska et al., 1978b
14 Eclogite (Col. Plateau)	8	3.28-3.53	7.85-8.41	8.09	0.1-11.3	3.7	0.5	Kumazawa et al., 1971
15 Eclogite	4	3.42-3.72	8.00-8.59	8.21	0.8-2.7	1.6	0.5	Kumazawa et al., 1971

composition in which pyroxene (omphacite) and garnet dominate. In this section several examples of elastic anisotropies of upper mantle rocks are given and then the results of different authors are summarized.

The Twin Sisters dunite was originally studied by Birch (1960a, 1961), who was the first to point out its extremely high elastic anisotropy caused by the preferred orientation of olivine grains. Christensen (1966b) demonstrated the importance of elastic anisotropy for ultramafic rocks and accomplished a study on elasticity of two samples of this dunite in correlation with their olivine orientations (Christensen and Ramananantoandro, 1971). The anisotropic elasticity of both samples was characterized in terms of the elastic constants of hexagonal and orthorhombic materials. A detailed investigation of the dunite anisotropy (Babuska, 1972b) showed a strong preferred orientation of olivine grains, resulting in P-velocity anisotropy of 15% at 1 GPa (Tab. 3-5).

The fabric diagrams of the Twin Sisters dunite (Babuska, 1972b) show distinct concentrations of all three olivine axes (Fig. 3-11). The direction of the highest

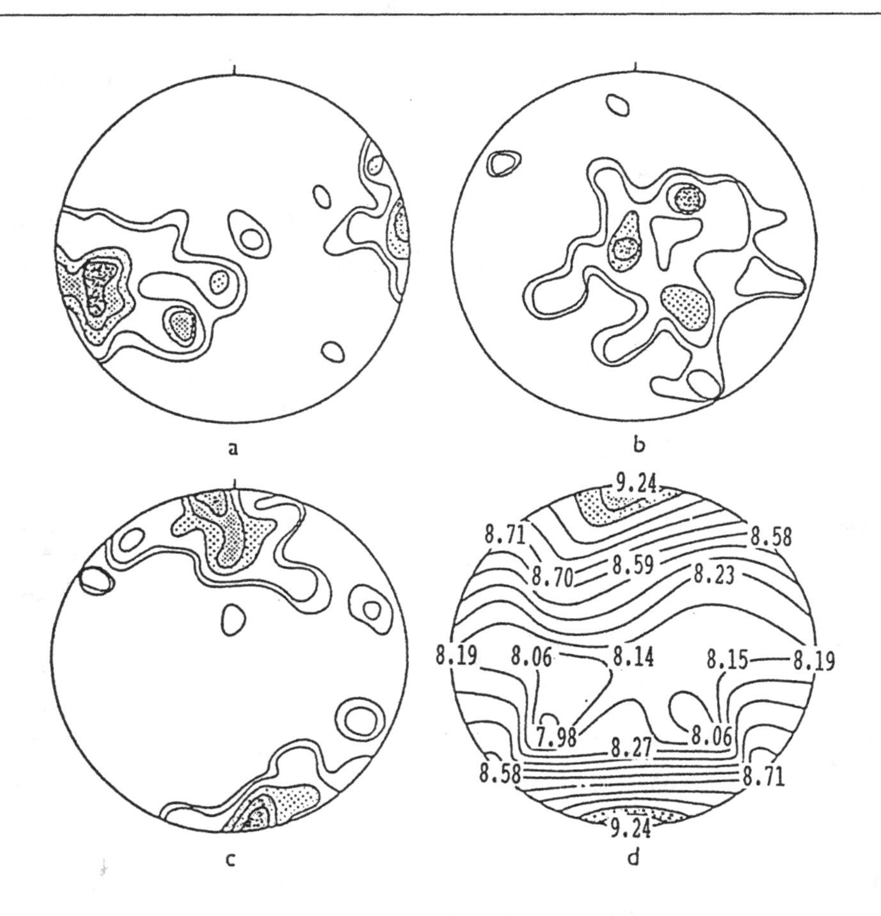

Figure 3-11 Orientation diagrams of 147 olivine grains in the Twin Sisters dunite and the diagram of elastic anisotropy (Babuska, 1972b). (a) b-axes, contours 8,6,4,2 and 1% per 1% area; (b) c-axes, contours 6,4,2 and 1% per 1% area; (c) a-axes, contours 8,6,4,2 and 1% per 1% area; (d) diagram of anisotropy showing the spatial distribution of compressional wave velocities; isolines are constructed from 8 to 9.2 km/s in steps of 0.1 km/s. (AGU, copyright © 1972).

P velocity is parallel to the maximum of olivine a-axes, the concentration of b-axes lies close to the direction of the lowest velocity. This observation corresponds to ultrasonic velocities in a single crystal of olivine (Kumazawa and Anderson, 1969). The plane of b and c-axes concentrations in the dunite is the plane of the lowest velocities of compressional waves. The shear wave velocities with directions of propagation and directions of displacement lying in this plane vary only insignificantly (from 4.57 to 4.63 km/s, see Fig. 3-12). On the other hand, the greatest differences of the S-wave velocities are found between the velocities of the S waves polarized in the plane of olivine b- and c-axes concentrations and the S waves polarized perpendicularly to this plane. It was found that orthorhombic symmetry is a reasonable approximation for the elastic anisotropy of the Twin Sisters dunite; with one exception, the deviations of the individual constants, determined from different waves, do not exceed a few per cent (Babuska and Sileny, 1981).

Lherzolites are characterized by lower anisotropy than dunite, mainly due to a higher content of pyroxenes. The anisotropy coefficients vary between 4 and 12%, with a mean value about 8%, at the confining pressure of 1000 MPa (Babuska, 1972a). The samples were xenoliths from volcanic eruptions. The average of the highest compressional velocities in the six investigated lherzolite xenoliths was 8.36 km/s, the average of the lowest velocities was 7.76 km/s. An example of the strong preferred olivine orientations and the corresponding anisotropy is shown in Fig. 3-10. The

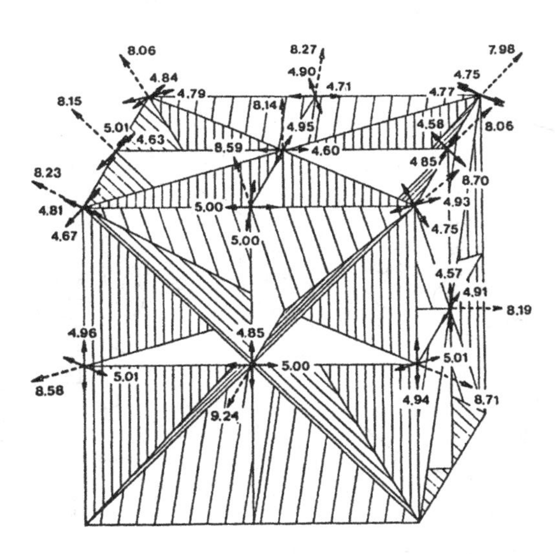

Figure 3-12 Schematic representation of the system of measurements in the Twin Sisters dunite (Babuska, 1972b) showing the values of ultrasonic velocities, directions of P- and S-wave propagation and directions of S-wave displacement. (AGU, copyright © 1972).

lherzolite of San Bernardino (California) is typical by strong lineations expressed by the maximum of compressional wave velocities corresponding to the maximum of olivine a-axes.

As the shear wave anisotropy provides very important information on the structure of the Earth's crust and upper mantle we present in Table 3-6 variations of shear-wave velocities and their anisotropy in ultramafic olivine-rich rocks determined at hydrostatic pressure of 1000 MPa by different authors. The values of the anisotropy coefficient k

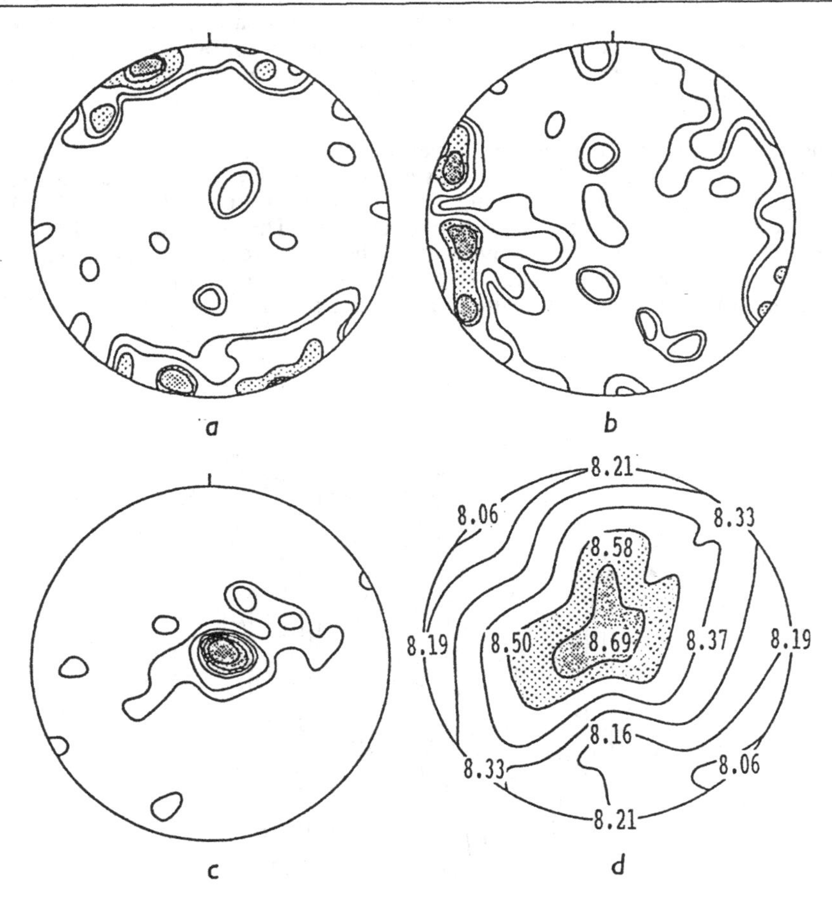

Figure 3-13 Lherzolite SB1 from San Bernardino, California (Babuska, 1972a), the equal-area projections of 150 olivine grains and the diagram of elastic anisotropy: (a) b-axes, contours 8,6,4,2 and 1% per 1% area; (b) c-axes, contours 6,4,2 and 1% per 1% area; (c) a-axes, contours 12,10,8,4 and 2% per 1% area; diagram of elastic anisotropy, isolines of compressional velocities are constructed from 8.1 to 8.6 km/s in steps of 0 .1 km/s. (With permission of Springer International, copyright © 1972).

also depend on the number of measurement directions. This means that a real anisotropy could be higher if the directions of velocity measurements were not chosen properly in accordance with the rock fabric. In general, the coefficient of anisotropy for S-wave velocities is lower than 10% and also seems to be lower than the coefficient of anisotropy for the P-wave velocities in olivine-bearing rocks. In addition to the mineral preferred orientation, also other structural phenomena play an important role in a large-scale seismic anisotropy, as it will be discussed in the next chapter.

Christensen and Crosson (1968) concluded that some ultramafic rocks showed a tendency toward olivine orientation, in which the crystallographic b-axes are concentrated normal to foliation and the a- and c-axes form a girdle in the foliation plane. These fabrics, which are typical for axially deformed specimens, behave macroscopically as transversely isotropic elastic solids with the lowest P-wave velocity perpendicular to the foliation plane. Another orientation pattern of olivine common in natural peridotites, which would produce transverse isotropy, but with a high P velocity parallel to the symmetry axis (and probably also to the horizontal flow directions) is characterized by concentrations of olivine a-axes and girdles of b- and c-axes perpendicular to that direction. Contrary to the first type, this orientation would cause azimuthal variations of Pn velocities, as observed in seismic refraction studies (e.g. Bamford, 1973).

Pyroxenites generally have lower anisotropy coefficients than olivine-rich rocks (Babuska, 1984). This is mainly caused by a lower degree of the mineral preferred orientations which are more random than those of dunite or lherzolite. Several pyroxenites representing xenoliths from volcanic rocks exhibit anisotropies between 1% and 4% (Babuska, 1972a).

Table 3-6 Shear-wave velocities and coefficients of anisotropy (k) in olivine ultramafic rocks.

Rock type	V_S^{max} [km/s]	V_S^{min} [km/s]	aver. V_S [km/s]	k%	Source of data
Dunite (Webster)	4.44	4.38	4.40	1.4	
Dunite (Mt. Dun)	4.64	4.43	4.54	4.6	
Dunite (T. Sisters)	4.90	4.70	4.83	4.1	
Dunite (Transvaal)	3.98	3.75	3.90	5.9	Simmons, 1964
Dunite A (T. Sisters)	4.696	4.362	4.529	7.4	Christensen and
Dunite B (T. Sisters)	4.980	4.747	4.864	4.8	Ramananantoandro, 1971
Peridotite 1 (Hawaii)	4.62	4.29	4.49	7.3	
Peridotite 2 (Hawaii)	4.75	4.62	4.68	2.8	
Dunite (T. Sisters)	4.89	4,51	4.74	8.0	Christensen, 1966b
Dunite (T. Sisters)	5.01	4.57	4.79	9.2	Babuska, 1972b
Lherzolite KH2 (N.Mex.)	4.85	4.42	4.63	9.2	
Lherzolite SB1 (Calif.)	4.89	4.61	4.75	5.9	
Lherzolite Ba2 (Calif.)	4.86	4.53	4.69	7.0	Babuska, 1976

The primary composition of eclogites is very simple: clinopyroxene (omphacite) and garnet are the main constituents. Especially those eclogites associated with ultramafic rocks are regarded as likely to have had a deep-seated origin. Birch (1960a) observed that although the velocities of eclogite and olivine-rich ultramafic rocks are comparable, the velocity anisotropy of eclogites is much smaller, and other investigations have suggested that this dichotomy may be diagnostic in deciding which rock type predominates in the upper mantle (Kumazawa, 1963; Bayuk et al., 1967).

In eclogite rocks from the Bohemian Massif both P- and S-wave velocities increase with the content of garnet and decrease with the content of symplectite, which originates in retrograde metamorphism of the rocks. The velocity anisotropy for both P and S waves is less than 4% at high hydrostatic pressure (Babuska et al, 1978b).

In general, the variability of the rocks, which are the most probable materials of the uppermost mantle (Table 3-5), both as to their composition and their physical parameters, is much lower than that of the crustal rocks (Table 3-4). Fig. 3-14 shows variations of P velocities due to compositional changes and due to anisotropy. The well-known high elastic anisotropy of peridotites holds for all olivine types, showing higher variations of velocities due to anisotropy than due to compositional changes.

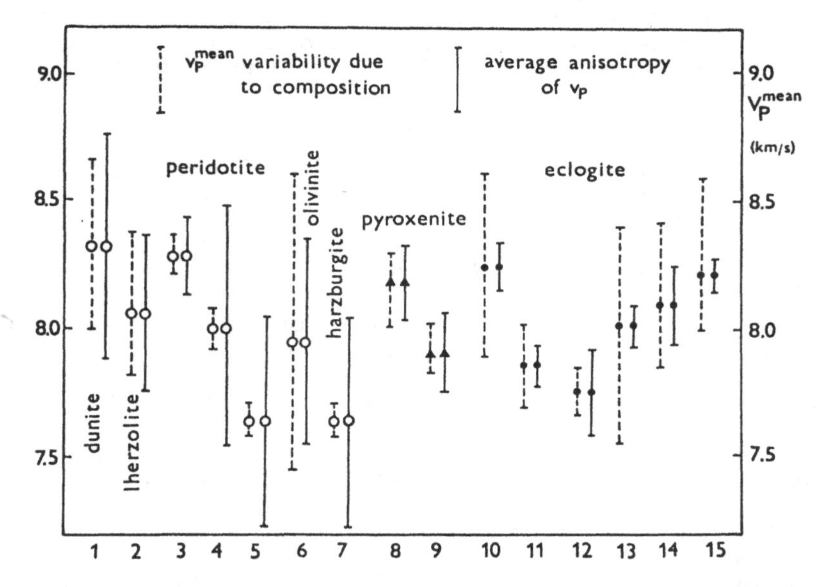

Figure 3-14 Mean velocities of compressional waves determined under high hydrostatic pressure in upper mantle rocks - olivine types (open circles), pyroxenites (solid triangles) and eclogites (solid circles). Numbers on the horizontal axis correspond to groups of samples in Table 3-4 (Babuska, 1984). (With permission of the Royal Astronomical Society, copyright © 1984).

Chapter 4 - Interpretation of seismic data in terms of anisotropy

A lot of seismic data present features which remain unexplained or poorly explained after an interpretation is made in terms of classical isotropic models. Seismologists are used to employing the word "noise" to explain why their models do not fit the data better. Even if part of the misfit is recognised as of deterministic nature, it is often small enough to be considered as due to some second order effects which are not taken into account in the theory. The problem is that these "second order effects" are sometimes not so small... It is then worthwhile to have a closer look at the data, be they parametric, such as arrival times or dispersion curves, or non-parametric, such as waveforms or amplitude spectra. Let us take the example of teleseismic P-wave arrival times measured on dense networks of seismological stations. Papers published on tomographic inversions of such parametric data quite commonly explain less than 50% of the initial variance of the observations. Very small variance reductions were indeed quite common in the earliest large-scale tomographic studies based on bulletin data. In Europe for example, the three-dimensional P-velocity model of Hovland and Husebye (1981) explained only 23% of the initial data variance while Romanowicz (1980) explained 33% of the variance for a similar data set. Errors in the readings of the arrival-times are often cited as the main source of noise in such data. In fact, even if the readings are not taken from the station bulletins but are controlled on the records themselves, the unexplained variance in the data can remain quite large.

Let us take a recent example of tomographic study in the upper Rhinegraben (Achauer et al., 1989). For this study, 60 portable stations were operated during six months in an area covering about 150×200 km^2 in order to record a large set of teleseismic P-wave arrivals. Cross-correlation techniques were used to eliminate the errors linked to the picking of the arrival times. Dense geophysical studies in this area, in particular high quality vertical seismic profiling made across the upper Rhinegraben, enable the authors to carefully correct the data for crustal effects before inverting them in terms of isotropic three-dimensional P-velocity models of the upper mantle. The best model obtained for the lithosphere and the asthenosphere explains 46% of the initial variance in the data (Glahn and Granet, 1991). By the standards of teleseismic tomographic studies, this score can be considered as a good one, giving some confidence in the three-dimensional image of the upper-mantle obtained in this region. There are, however, more than 50% of the initial variance of the data which remains unexplained by the isotropic three-dimensional models. What is the cause of this misfit to the data? Is it just noise in arrival times? Is it due to second order effects not taken into account in the isotropic model such as inadequate ray tracing or non-linear effects in the inverse theory, or is it just anisotropy present somewhere along the ray paths?

In other branches of seismology, there are seismic models fitting the data fairly well but which may be quite unrealistic if some pertinent physical or mineralogical data are considered. For example, when searching for the source parameters of the Imperial Valley earthquake of October 15th, 1979, complicated models of the source process were proposed to explain the observed polarization of strong motion records at short epicentral distances (Liu and Helmberger, 1985; Gariel et al., 1990). A much simpler source process can indeed explain the data if some anisotropy is put in the upper crust, causing S-wave splitting on the observed strong-motion records (Zollo and Bernard, 1989). In a completely different field, long-range seismic profiles sometimes necessitate very high seismic velocities in the lithosphere to fit the refraction data in the framework of an isotropic model. As these seismic velocities may be difficult to explain with the known mineralogy of upper mantle rocks, why not invoke anisotropy as one possibility since we know that overall anisotropy can be present within realistic upper mantle materials when they are deformed by a coherent large-scale strain? In yet another field, surface-wave investigations of the upper mantle sometimes necessitate a very low velocity zone in the asthenosphere which may be unrealistic in view of our knowledge of the mineralogy and the physical conditions at a depth of 100 or 200 km (Anderson and Regan, 1983). Can anisotropic modelling change the structure of the low-velocity zone which has been inferred from surface-wave data within an isotropic framework? We will come back to these different examples later on, but let us first address in very general terms the basic question of a trade-off between anisotropy and heterogeneity, indeed a key question for any seismic study of complex media, which lies behind all the above interrogations.

4.1 Trade-off between seismic anisotropy and heterogeneity

Seismic-wave propagation is affected by both heterogeneity and anisotropy of the elastic medium where the propagation takes place. We have already stated that ignoring anisotropy may lead to erroneous conclusions on the physical properties of the Earth's interior. Ignoring heterogeneities as the source of a possible explanation of observed anomalies in the seismic data, can also lead to wrong images of the Earth's interior. Such a possibility was pointed out for example by Levshin and Ratnikova (1984) in a short note which we think interesting to examine here in detail. These authors took two very simple examples to illustrate how isotropic models presenting heterogeneities can explain some anomalous behavior of the seismic wavefield which is otherwise commonly explained in terms of anisotropy. The first example is the classical problem of the stack of isotropic layers which we have already discussed in detail in section 2.1: there is obviously no way to distinguish intrinsic anisotropy on the scale of the crystal from fine layering of isotropic materials if the layer thicknesses are much smaller than the seismic wavelengths. There is a complete trade-off in this case between homogeneous anisotropic models and heterogeneous isotropic models and there is no way to discriminate between them from long wavelength seismological observations alone. This problem can thus be considered mainly as of semantic nature: we have to use just adequate wording to design such large-scale, shape-induced anisotropy, in intrinsically isotropic materials. In this monograph, we use the term "seismic anisotropy" for this kind of overall anisotropy as well as for the intrinsic anisotropy.

The second example examined by Levshin and Ratnikova (1984) is more relevant to the questions discussed in this chapter. It concerns one of the observations which is usually considered as a non-ambiguous sign of seismic anisotropy: the Love/Rayleigh -wave discrepancy in surface-wave seismology. In this example, two-layered isotropic and elastic quarter spaces in a welded contact along a vertical boundary are considered. These authors have performed a simple numerical experiment by computing linearly the average of the phase and group slownesses (inverses of velocities) for Love and Rayleigh waves propagating from point A to point B, crossing the vertical plane boundary at right angle (Fig. 4-1).

Let d_1 and d_2 be the path lengths in the first and second quarter spaces, and C_n (T) and U_n (T) be the phase and group velocities at period T in the quarter space n, respectively. The average phase and group velocities along the path AB are simply given by:

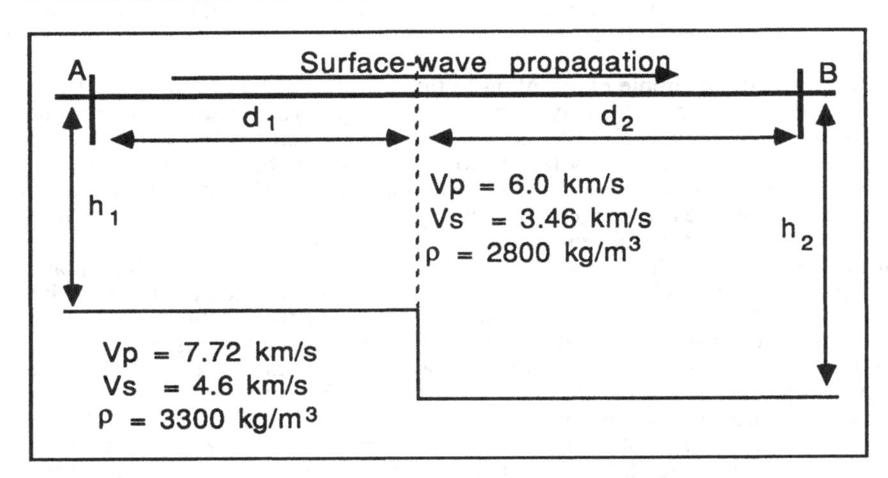

Figure 4-1 Isotropic laterally heterogeneous model considered by Levshin and Ratnikova (1984) as a source of apparent anisotropic behavior of surface waves (i.e. Love/Rayleigh-wave incompatibility which is not due to intrinsic anisotropy). The thickness of the upper layer varies from h_1=35 km to h_2=50 km .

$$C(T) = (d_1 + d_2) / (d_1 / C_1(T) + d_2 / C_2(T)),$$

and

$$U(T) = (d_1 + d_2) / (d_1 / U_1(T) + d_2 / U_2(T)).$$

No phase shift was introduced at the boundary between the two quarter spaces as is usual in such computations dealing with the main seismic wavefield of the fundamental mode of the Love and Rayleigh waves.

The average dispersion curves C(T) and U(T) of the Love and Rayleigh waves were then simultaneously inverted in terms of laterally homogeneous and isotropic models using a non-linear inversion scheme, the hedgehog technique (Keilis Borok et al., 1969). In this technique the multi-dimensional space of the model parameters is statistically explored, starting from an approximate solution. Exact predictions of phase and group velocities are then compared with the observation. The models presenting the best fit to the data are kept as the most likely models. The result of this numerical experiment is that no single laterally homogeneous isotropic model can explain simultaneously the Love- and the Rayleigh-wave average dispersion curves and separate inversions of the Love- and Rayleigh-wave dispersion curves. This means that some "Love/Rayleigh-wave discrepancy" has been created which is not due to anisotropy. As only direct computations are performed in this noise-free numerical experiment, we cannot invoke artifacts due to the inverse method. The inability to fit simultaneously the Love and Rayleigh curves with an average laterally homogenous model seems to be real and another explanation has to be found.

We propose here a simple explanation to this puzzling result. First of all, let us stress the fact that there is necessarily some non-linearity involved somewhere in the numerical experiment. Indeed, if the differences between the velocity models of the two layered quarter spaces were linearly related to the differences between the phase and group velocities, $C_1(T) - C_2(T)$ and $U_1(T) - U_2(T)$, everything should add linearly in this experiment. The linear average of the two quarter space models should explain perfectly the average dispersion curves and no discrepancy should then show up between the Love- and Rayleigh-wave dispersion curves. As such a model based on a linearity hypothesis does not give the best fit to the dispersion data, one has to seek a non-linear effect. It is very likely that it is the relationship between the surface-wave dispersion curves and the seismic velocities in the quarter-space models, which is the source of non-linearity. In Levshin and Ratnikova's (1984) numerical experiment, one quarter space represents a normal continental crust with a 35-km deep M discontinuity while the other quarter space represents a 50-km thick continental crust. Such lateral heterogeneities are indeed not unrealistic but, in terms of the lateral variation of velocities in the depth range 35-50 km, it is quite large, the velocity contrast attaining 33%. It is likely that this lateral variation of seismic velocity exceeds the limit of linearity of the relationship relating the velocity model to the dispersion curves, mainly for group velocities at the shorter periods, so that the Love waves do not see the same horizontally averaged depth-velocity model as the Rayleigh waves do. Since no single velocity model could then fit the two sets of Love- and Rayleigh-dispersion curves, this could explain why an apparent Love/Rayleigh-wave discrepancy shows up in the experiment.

If the above explanation is correct, Levshin and Ratnikova's experiment shows what can be the limitations of surface-wave investigations of anisotropy based on observations of Love/Rayleigh-wave discrepancies: it is clear that classical linear inversion schemes of surface-wave dispersion curves should not be applied to regions of too strong lateral heterogeneities without special attention being given to this problem. Let us note, however, that the largest lateral heterogeneities of seismic velocity are generally linked with the variation of the M-discontinuity depth, so that a careful correction for the effects of the lateral variation of the crustal layer thicknesses on the average dispersion curves can reduce the risk of a misleading interpretation of an observed Love/Rayleigh-wave incompatibility.

Let us now take an example of trade-off between anisotropy and heterogeneities in a completely different field. Teleseismic-wave arrivals observed on dense networks of seismological stations are now traditionally interpreted in terms of three-dimensional seismic velocity structures. Perturbations to the travel time due to the three-dimensional heterogeneities are generally considered as small linear perturbations, only depending on the velocity along the ray path (Aki et al., 1977). Let us illustrate here this technique by taking a very simple example: a two-dimensional model of seismic velocities with two data related to the travel times along two ray paths and two unknowns related to two blocks of an isotropic velocity model. This model is displayed in Fig. 4-2a. The travel time along the first ray is:

$$t_1 = W / V_1,$$

where W is the width of the blocks and V_1 is the seismic velocity in block 1. The travel time along the second ray is

$$t_2 = L_1 / V_1 + L_2 / V_2,$$

where L_1 and L_2 are the lengths of blocks 1 and 2, V_2 is the seismic velocity in block 2. If the boundaries of the blocks are known, the inverse problem is a trivial linear

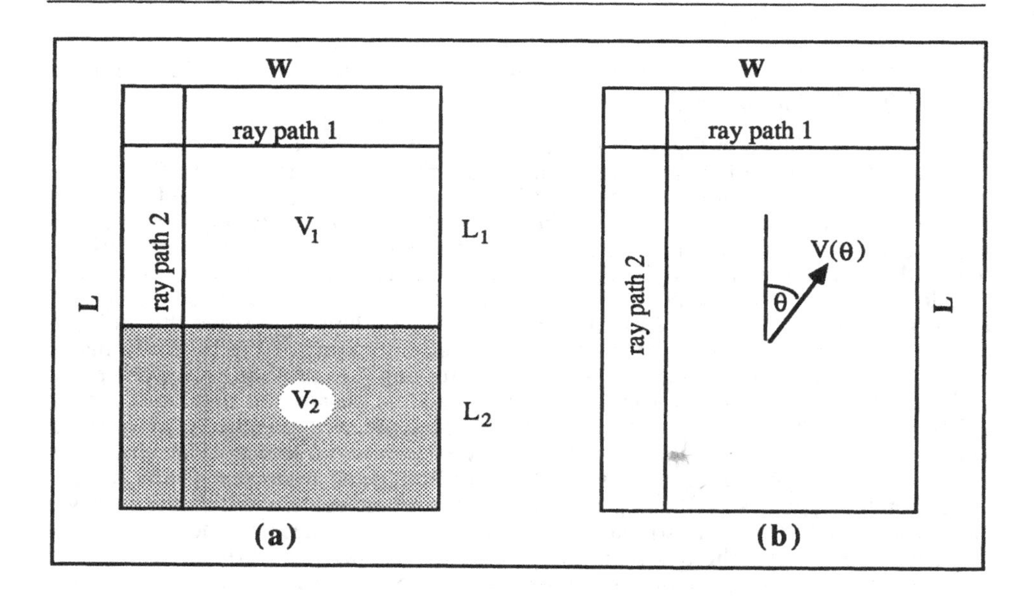

Figure 4-2 Simple two-dimensional velocity model illustrating the type of inverse problems solved in tomographic studies. (a) corresponds to a two-block isotropic model and (b) to a homogeneous anisotropic model.

system of two equations with 2 unknowns; and it is easy to retrieve velocities V_1 and V_2 from data t_1 and t_2. If nothing is known about the boundaries between the blocks, or if the seismic velocities vary continuously within the model, there is of course an infinite number of isotropic velocity distributions which fit the data. The resolution analysis would show however, as is usual in any linear inversion scheme, that averages of velocities can be obtained over spatial dimensions not very different from the size of the blocks drawn in Fig. 4-2a.

Let us now consider the homogeneous anisotropic model shown in Fig. 4-2b. In this case we have introduced an azimuthal variation of velocity $V(\theta)$ as follows:

$$V(\theta) = V_0 + \delta V \cos(2\theta + \theta_0).$$

This model can fit the observation of the two travel times t_1 and t_2 as well as the two-block isotropic model. It is even simpler in the sense that, now, we have a homogeneous model so that there is no block geometry to deal with. If θ_0 is known, we can solve exactly this second problem with two unknowns V_0 and δV and two data t_1 and t_2. Explicitly, stating for example $\theta_0 = 0$, one has:

$$V_0 = (W / t_1 + L / t_2) / 2,$$

$$\delta V = (L / t_2 - W / t_1) / 2.$$

If θ_0 is not known, more data are needed to find the solution of the problem with the three unknowns θ_0, V_0, and δV.

Models displayed in Fig. 4-2a,b can fit equally well the two data t_1 and t_2. If only these two data are available, we are in the presence of a complete trade-off between heterogeneity and anisotropy. Generalizing to larger three-dimensional models, one can foresee that separating anisotropy from lateral heterogeneity in tomographic investigations is a difficult, if not impossible, problem when no *a priori* constraints are put on the model. One can argue that such a trade-off is just an artifact due to the too simple model considered and that increasing the number of rays crisscrossing the investigated area, and thus the number of data, would allow us to solve the problem easily. It is clear that if one divides the rectangular domains of Fig. 4-2 into many elementary cells, and if a large number of crisscrossing rays provide each elementary cell with a good azimuthal coverage, it would become possible to resolve the local azimuthal variation of velocity by assuming, for example, a homogeneous anisotropic velocity within each of the elementary cells. Two-dimensional tomographic studies of surface-wave velocities proceed this way (e.g., Tanimoto and Anderson, 1985). In fact, if we drop the hypothesis of homogeneity of the anisotropic velocity within the cells, we can always subdivide them, so that only a few rays will hit these smaller cells. For a finite number of rays, there is thus always a scale on which an intrinsic problem of trade-off between anisotropy and lateral heterogeneity occurs: cells which would be well resolved in an isotropic investigation may be no more resolved if local azimuthal variation of velocity is allowed for.

Resolving local anisotropy from a finite set of travel-time data thus appears as an ill-posed problem. There is a need for some *a priori* information to be put on the spatial distribution of the model parameters if one wants to interpret the travel time data in terms of a smoothly varying heterogeneous and anisotropic model. Three types of *a priori* information can be put in the model for this purpose: 1) bounds on the local velocity distributions, 2) spatial correlation lengths for the physical parameters, and 3) bounds on the local azimuthal variations of velocities. If the *a priori* correlation length is large enough and if many crisscrossing rays affect an area of the size of this correlation length, it is clear that the azimuthal variation of velocity can be resolved locally. If the correlation length is too small, complete trade-off between the spatial variation of velocity and its azimuthal variation will occur. A worse situation to fit the data in a tomographic experiment would be to allow only for small and smooth spatial variations of velocity (i.e. small *a priori* bounds on velocity and large correlation lengths), without allowing for anisotropy (small *a priori* bound on the local azimuthal variations). Such isotropic *a priori* constraints on a three-dimensional structure are not well suited to fit travel-time data if there are large non-random spatial velocity fluctuations on a small scale or if the structure is intrinsically anisotropic. Such a choice, however, is quite commonly made in classical tomographic studies, the departure from a laterally homogeneous isotropic model being generally considered as small, smooth and isotropic.

Separating the effects of seismic anisotropy from those linked with lateral heterogeneity is thus by no means an obvious problem. It may even be an ill-posed problem with no solution if the scale of interest is set too small. Particular attention should thus be paid to casting this problem in the most objective way possible. This is obviously fairly easy in a testing-hypothesis approach where the predictions from different theoretical models are compared with the data in order to validate or invalidate them. This is much more difficult when we attempt to perform an objective inversion of the seismic data. Several mathematical tools have been developed during the last two decades to help solve this question as a multiparameter inverse problem. Indeed, inverse methods are certainly the only way to base on solid grounds the discussions on the trade-off between seismic anisotropy and heterogeneity.

4.2 Seismic anisotropy and inverse methods

Inverse methods have been designed to interpret a set of data in terms of model parameters in the most objective way possible (e.g. Tarantola, 1987). The linear inverse theory is well advanced. Algorithms based on linear programming methods (Sabatier, 1977) and on least square techniques (Backus and Gilbert, 1968, 1970; Franklin, 1970; Wiggins, 1972; Jackson, 1972) are now classical and are applied to a wide set of geophysical data. In the framework of a very general stochastic approach, Tarantola and Valette (1982) have extended the least square techniques to weakly non-linear systems by considering Gaussian statistics of the model parameters and the data. In the discrete version of their algorithm, the model is a finite set of parameters which can be represented as a vector $\mathbf{m}$. At iteration k+1 the inverted model $\mathbf{m}_{k+1}$ minimizing the differences between the data vector $\mathbf{d}$ and the predicted data on the one hand, and the differences between the inverted model $\mathbf{m}_{k+1}$ and the *a priori* model $\mathbf{m}_o$, scaled by their respective *a priori* covariance matrices $\mathbf{C}_{do}$ and $\mathbf{C}_{mo}$ on the other hand, is given by

$$\mathbf{m}_{k+1} = \mathbf{m}_o + \mathbf{C}_{mo}\, \mathbf{G}_k^t\, (\mathbf{C}_{do} + \mathbf{G}_k\, \mathbf{C}_{mo}\, \mathbf{G}_k^t)^{-1}\, [\mathbf{d} - \mathbf{d}(\mathbf{m}_k) + \mathbf{G}_k(\mathbf{m}_k - \mathbf{m}_o)],$$

where $\mathbf{m}_1$ is the starting model usually taken equal to $\mathbf{m}_o$, $\mathbf{d}(\mathbf{m}_k)$ is the column data vector predicted from the model $\mathbf{m}_k$, $\mathbf{G}_k$ is the matrix made of the partial derivatives of the observations with respect to the model parameters at step k, and upper index t denotes the matrix transposition.

If the direct problem relating the model to the data is linear, or if it can be reasonably linearised in the vicinity of the *a priori* model $\mathbf{m}_o$, the inversion algorithm is reduced to a one-step procedure. The equation relating a model perturbation $\delta\mathbf{m}=\mathbf{m}-\mathbf{m}_0$ to the predicted data $\delta\mathbf{d}$ is then simply

$$\delta\mathbf{d} = \mathbf{G}\, \delta\mathbf{m},$$

and the equation relating the data vector **d** to the inverted model $\mathbf{m^*}$ becomes

$$\mathbf{m^*} = \mathbf{m_o} + \mathbf{H}\,[\mathbf{d} - \mathbf{d}(\mathbf{m_0})]$$

where the inverse operator **H** is:

$$\mathbf{C_{mo}}\,\mathbf{G^t}\,(\mathbf{C_{do}} + \mathbf{G}\,\mathbf{C_{mo}}\,\mathbf{G^t})^{-1}\,.$$

This equation was derived for the first time by Franklin (1970) within the framework of a stochastic approach. In this algorithm, the basic trade-off between resolution and errors of the model parameters (Backus and Gilbert, 1970) is handled by playing with the values of the diagonal terms of the *a priori* covariance matrices $\mathbf{C_{do}}$ and $\mathbf{C_{mo}}$ (Jackson, 1979).

Resolution of the model parameters is given by the matrix

$$\mathbf{R} = \mathbf{HG}$$

or

$$\mathbf{R} = \mathbf{C_{mo}}\,\mathbf{G^t}\,(\mathbf{C_{do}} + \mathbf{G}\,\mathbf{C_{mo}}\,\mathbf{G^t})^{-1}\mathbf{G}.$$

This resolution matrix relates any perturbation $\boldsymbol{\delta}\mathbf{m}$ of the *a priori* model which fits the data according to the *a priori* covariance matrices $\mathbf{C_{do}}$ and $\mathbf{C_{mo}}$, to the inverted model perturbation $\boldsymbol{\delta}\mathbf{m^*} = \mathbf{m^*}\text{-}\mathbf{m_o}$:

$$\delta\,\mathbf{m^*} = \mathbf{R}\,\boldsymbol{\delta}\mathbf{m}.$$

Note that although the resolution matrix **R** depends on the *a priori* covariance of the data, it does not depend on the values taken by the data **d**. Inferences on the resolution can thus be done whatever the actual values of the data may be. A perfect resolution would mean that **R** is a unit diagonal matrix so that the inverted model $\mathbf{m^*}$ equals the actual model **m**. The trade-off between the model parameters can be analysed by examining the off-diagonal terms of **R**. Note that the resolution matrix is not diagonal so that the trade-off properties are not symmetric. Note also that the inverted model $\mathbf{m^*}$ is one of the possible candidates for the actual model. It fits the data in agreement with the *a priori* covariance matrices of the model parameters $\mathbf{C_{mo}}$ and the data $\mathbf{C_{do}}$ as well as the real Earth's model, so that

$$\delta\,\mathbf{m^*} = \mathbf{R}\,\delta\,\mathbf{m^*}.$$

Most of the basic questions related to the inversion of seismic data in terms of anisotropic parameters can be analysed within the framework of the linear least-squares method so that hereafter we will only discuss within such a framework the questions related to anisotropic inversion of seismic data. The trade-off between anisotropy and heterogeneity in particular, can be analysed by looking at the resolution matrix **R**. If an anisotropic tomographic problem can be linearized with some good degree of confidence, one line of the resolution matrix gives the trade-off between one anisotropic parameter, for example the amplitude of azimuthal variation of velocity in one block, and the other "isotropic" parameters of the model, e.g. the velocities of the blocks.

Inversely, just by examining other lines of the resolution matrix, it is possible to foresee whether the local azimuthal variation of velocity in the different parts of the model can affect the seismic velocity of a given block. If so, trade-off exists in the inverted tomographic image between anisotropy and heterogeneity. Such a resolution analysis is not commonly performed in tomographic investigations taking into account a large number of data and model parameters due to numerical difficulties in estimating the matrix **R**. In fact, large tomographic problems do not rely on simple matrix methods such as the one discussed above, due to the very large size of the matrices to be handled. Row action methods such as those used in medical tomography are often applied (Van der Sluis and Van der Vorst, 1987; Humphreys and Clayton, 1988). Nevertheless, Trampert and Leveque (1990) have noticed that for any linear inverse problem it is always possible to estimate the resolution matrix **R** by running the inverse operator **H** as many times as there are model parameters, i.e., the column i of the resolution matrix $\mathbf{R}_i$ equals the product $\mathbf{HG}_i$ if $\mathbf{G}_i$ is the column i of **G**.

After these general comments, let us now look at the specific problems linked with the anisotropy in inversion theory. Anisotropy is a source of major complications in any inversion of seismic data since the Earth models cannot any longer be treated as a single scalar functional describing the spatial variations of the seismic velocities. We have seen that, even with systems presenting locally a cylindrical symmetry, five parameters were needed to describe completely the elastic properties. If we note that two angles are necessary to orient the axis of symmetry and that density is one additional parameter to be considered in any modeling of the seismic signal, we are faced with eight parameters instead of three in the isotropic case. In other symmetry systems of interest for geophysical investigations, we are even left with more parameters to handle simultaneously. The major question is thus twofold: first we have to deal with a large increase of model parameters when passing from an isotropic approach to an anisotropic one, and second, it is necessary to take properly into account the differences in the physical units of the model parameters.

Before looking at the latter problem, let us again stress the fact that the best way to prepare an anisotropic inversion of seismic data is first to reduce as much as possible the number of model parameters. If some simplifying assumption can be made on the physical cause of anisotropy, it becomes indeed possible to reduce drastically this number of parameters. For teleseismic waves, one such possibility is offered by the presence of a rich-olivine content in the upper mantle, the very likely candidate for most of the observed seismic anisotropy of mantle waves. According to compositional estimates of the upper mantle materials and making some simplifying assumptions, we can restrict the anisotropic parameters to a percentage of oriented anisotropic crystals and to angles of the preferred orientation directions of the crystals in an otherwise isotropic matrix. A minimum of two angles are necessary to orient the olivine crystals statistically if, after Kirkwood and Crampin (1981), it can be assumed that only the a-axes of the olivine crystals are preferentially oriented in one direction while the two other crystallographic axes are randomly oriented in the perpendicular plane (see appendix 1). Such a hexagonal-symmetry model of olivine assemblage presents seismic properties which are not drastically different from the actual orthorhombic β-spinel crystals (Kirkwood and Crampin, 1981). With such a model, we are left with three parameters plus one isotropic velocity, either P or S velocity according to the type of seismic data. Four parameters are thus the miminum number of parameters which can be used to describing the seismic velocities in an olivine-rich upper mantle (Cara et al., 1984). In their three-dimensional anisotropic modeling of the upper mantle beneath

Japan, Hirahara and Ishikawa (1984), for example, reduced the number of anisotropic parameters by assuming an olivine-rich upper-mantle composition, but these authors preferred using a more complete parameterization: instead of the four parameters proposed above, they considered six unknowns to model the complete spheroidal velocity surface of an orthorhombic olivine material plus one isotropic velocity. Finally, in order to reduce the size of the parameter space, it is also possible to add petrological constraints as proposed by Montagner and Anderson (1989a, b) for the upper mantle.

A second example of a reduction of anisotropic parameters based on physical grounds is offered by crack-induced anisotropy in the upper crust. If the cause for anisotropy is due to the preferred orientation of fluid-filled cracks and if the cracks can be considered as randomly oriented but with the normals to their surface statistically oriented in one direction, hexagonal symmetry is the simplest anisotropic system which can be used to describe the seismic anisotropy in this case. According to the formulation of Hudson (1980, 1981) for a small crack density with oblate spheroidal shapes, it is possible to consider a reduced set of parameters to describe the medium (cf. p. 13). If the radius and aspect ratio of the cracks are fixed, the crack density and the preferred orientation direction of the normals to the crack surfaces are enough to describe the overall anisotropic properties of the medium. Three anisotropic parameters are then sufficient in this case. If the plane of the cracks can be considered as vertical, one angle only is needed, so that the minimum number of parameters which are necessary to describe the upper crust can then be reduced to two in addition to the isotropic velocity. Again, strong physical assumptions are introduced in the model and more parameters would be needed to fit the reality more closely.

Any such type of parameterization of the Earth's materials belongs to some mixed approach, combining hypothesis testing for the fixed parameters and inversion for the free parameters. Several physical hypotheses can be tested in this way while inverting the reduced set of unknown parameters. Even if the result of an inversion is that no anisotropy is required, one can have much greater confidence in the isotropic model thus obtained if the inverse method is set in terms of an anisotropic framework, or, better, several anisotropic parameterizations. Considerations such as the quality of the data fit, or the plausibility of the values obtained on the inverted physical parameters can be used as criteria to select the more credible physical hypothesis regarding the problem under study. In such approaches, it is clear that very strong constraints are put on the physical models we are searching for and oversimplifying assumptions can be made, but let us stress the fact that the *a priori* constraints are even stronger when an isotropic formulation is used. Indeed, even if anisotropy is not obvious in the data, such a type of modeling is preferable since isotropy is always allowed for in the inversion when the coefficients of anisotropy become null.

Let us now look at the problem of physical inhomogeneity of the model parameters. In an anisotropic structure, it is no longer possible to seek the variation of a single physical parameter, such as the P velocity varying from block to block in a three-dimensional space in a tomographic investigation. No matter how many anisotropic parameters are reduced in the modeling, more than one kind of physical parameter has to be taken into account, such as seismic velocity and percentage of oriented cracks for example. The inversion of such an inhomogeneous set of parameters is the source of specific problems in the inverse theory, which are known as the consistency problem. It is indeed not obvious to state the inverse problem in such a way that the output of the inversion is consistent, i.e. independent of the specific set of parameters chosen to describe the physical model.

To illustrate this important question of consistency in inverse methods, let us take an example of the inversion of a set of surface-wave data from a study of dispersion curves presenting a Love/Rayleigh-wave discrepancy for both the fundamental modes and the higher modes (Leveque and Cara, 1985). The purpose of this study was to perform an inversion of a set of Love- and Rayleigh-wave dispersion curves, including higher mode branches up to the third overtone, in terms of a transversely isotropic structure of the upper mantle with a vertical symmetry axis. The inversion was made with the use of a classical singular-value decomposition of Lanczos (1961) according to the algorithm proposed by Wiggins (1972). Due to the large quantity of information available from the set of higher mode data which had been chosen in that study, it was actually an interesting exercise to test how many independent pieces of information were resolvable to constrain the models of upper-mantle anisotropy. For this experiment, contrary to the philosophy developed above for the mixed approach of testing hypothesis / inversion, we decided not to put any additional constraint on the inverted parameters. The assumption of a vertical axis of symmetry was already a strong enough constraint indeed. Five elastic coefficients plus the density were thus allowed to vary along the depth coordinate z in the inversion process. Explicitly, following the notations of Takeuchi and Saito (1972) and Love (1911), Leveque and Cara (1985) used the six following parameters:

$\alpha_V(z)$	P_V velocity,	km/s,
$\beta_V(z)$	S_V velocity,	km/s,
$\rho(z)$	density,	kg/m^3,
$\xi(z)$	$= N / L = [\, \beta_H(z) / \beta_V(z)\,]^2$	dimensionless,
$\varphi(z)$	$= C / A = [\, \alpha_V(z) / \alpha_H(z)\,]^2$	dimensionless,
$\eta(z)$	$= F / (A - 2L)$	dimensionless,

where α_V and α_H denote the velocity of P waves propagating horizontally and vertically, and β_V and β_H are the velocities of horizontally travelling S waves for horizontal and vertical polarization directions, respectively. In isotropic solids the anisotropic parameters are equal ($\xi = \varphi = \eta = 1$). Note that the above notation slightly differs from that introduced by Anderson (1961), η being defined as the inverse of η by Anderson's (1961).

For the purpose of this discrete inversion experiment, the above functions were sampled at 22 points along the depth coordinate. Note that α_V, β_V, and ρ are not in the same physical units as the three anisotropic parameters ξ, φ, and η, so that they were normalized to their maximum values in the involved depth interval. A first inversion was performed by considering β_V and ξ as free parameters. The result of the inversion, made by projecting the data on the 13 eigenvectors associated with the largest eigenvalues is shown in Fig. 4-3 for the ratio of S_H to S_V velocities versus depth (curve A). As a large Love/Rayleigh-wave discrepancy, with apparently too fast Love waves, was observed in the dispersion curves, it is not surprising to obtain larger S_H velocities than S_V velocities in the inverted model. The fact that in this numerical experiment large differences extend to a depth of more than 300 km does not correspond to reality because the specific set of data used for this figure is meaningless.

In order to test the sensitivity of the inverted model to the physical parameterization of anisotropy, the whole set of data was then re-inverted using the two velocities of S_H and S_V waves instead of one velocity and the square of their ratio $\xi(z)$ as independent parameters. This second inversion corresponds to the same physical hypothesis, i.e. a transversely isotropic model. It was made by projecting the data on the same number of eigenvectors. The result is shown in Fig. 4-3 (curve C). Drastic changes appear in the inverted model which does not give the same S_H to S_V velocity ratio as the former one. Inverting the data within the framework of a third parameterization of the elastic properties ($\beta_V(z)$, $\xi(z)^{1/2}$) still leads to a different model (curve B). This shows that the inverse method we have used is not consistent.

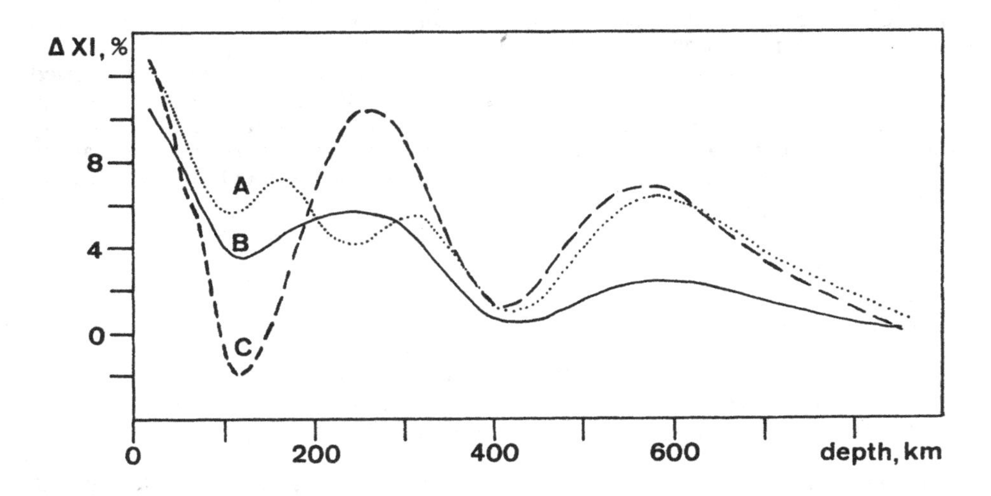

Figure 4-3 Example of inconsistent outputs of an inverse method: the three parameterizations of the anisotropic elastic properties used in the inversion process, $\{\beta_V(z), \xi(z)\}$ (A), $\{\beta_V(z), \xi(z)^{1/2}\}$ (B) and $\{\beta_V(z), \beta_H(z)\}$ (C), do not yield the same ratio $\xi(z) = (\beta_H/\beta_V)^2$ while the inversion is made with the same number of eigenvectors. The starting model for ξ is set equal to unity and the figure shows the perturbation $\delta\xi(z)$ in percentage so that $\xi = 1 + \delta\xi / 100$ (from Leveque and Cara, 1985). (With permission of the Royal Astronomical Society, copyright © 1985).

The reason for the observed inconsistencies lies in the hidden assumptions made on the *a priori* information which is implicitly admitted for the model parameters. In a classical application of the inversion algorithm of Wiggins (1972), it is implicitly assumed that the *a priori* covariance matrix of the model parameter is diagonal. The problem of consistency which is shown in Fig. 4-3 when we change the physical parameterization of the problem lies in the fact that if the *a priori* covariance matrix of the model parameters is diagonal in the parameter set $\{\beta_V(z), \beta_H(z)\}$, it is not diagonal in the set $\{\beta_V(z), \xi(z)\}$. Leveque and Cara (1985) have shown that although no such concept was introduced in the singular value decomposition algorithm of Wiggins (1972), it was easy to make the algorithm consistent by re-scaling the equation to an *a priori* covariance matrix and by fixing in an objective way the number of eigenvectors to be kept in the final model according to Matsu'ura and Hirata (1982). The idea of rescaling the equations to some *a priori* covariance matrix of the model parameters is not new. It was described in great detail by Jackson (1979). Such an idea was systematically applied by Tarantola and Valette (1982). Their method is consistent. The original algorithm of singular-value decomposition of Wiggins (1972) is not.

The way to make consistent the inversion results shown in Fig. 4-3 is to introduce non-diagonal terms in the *a priori* covariance matrix for two out of the three parameterizations. Considering small perturbations $\delta\beta_V(z)$ and $\delta\xi(z)$ of the model parameters, the matrix transformation from the set of parameters $\{\beta_V(z), \xi(z)\}$ to the set $\{\beta_V(z), \beta_H(z)\}$ is as follows:

$$\begin{pmatrix} \delta\beta_V \\ \delta\beta_H \end{pmatrix} = \begin{pmatrix} 1 & 0 \\ \xi^{1/2} & \dfrac{\beta_V}{2\xi^{1/2}} \end{pmatrix} \cdot \begin{pmatrix} \delta\beta_V \\ \delta\xi \end{pmatrix} .$$

Covariance matrices transform as $C_{x'}^{-1} = T^t C_x^{-1} T$, if T is the transformation matrix between the two sets of parameters **x** and **x'**. If the *a priori* covariance matrix is diagonal in the set of parameters $\{\beta_V(z), \xi(z)\}$, its expression in the set of parameter $\{\beta_V(z), \beta_H(z)\}$ is thus:

$$C_{x'} = \begin{pmatrix} \sigma_\beta & \sigma_\beta \\ \xi^{1/2}\sigma_\beta & \xi\sigma_\beta + \dfrac{\beta_V^2}{4\xi}\sigma_\xi \end{pmatrix}$$

where σ_β^2 and σ_ξ^2 are the variances of β_V and ξ.

A coherent setting of the *a priori* covariance matrices, as above, would thus have been necessary to make the inverted models consistent. To conclude these remarks on the consistency problem, let us emphasize the fact that the setting of the *a priori* covariance matrix of the model parameters is by no means a simple exercise, in particular if off-diagonal terms have to be defined. The reason is that we have to quantify our *a priori* knowledge on the parameters and its correlation with other parameters before an inversion of a new data set is performed. The way of quantifying this *a priori* knowledge is always somewhat subjective. One has first to decide in which parameterization the matrix is diagonal, if any. In the example presented above, we have assumed that our knowledge of the ratio of S_H to S_V velocities (set equal to unity before the inversion) was *a priori* uncorrelated with our knowledge of S_V velocities, so that this parameterization corresponds to a diagonal *a priori* covariance matrix. This assumption means implicitly that our knowledge of S_H and S_V velocities is *a priori* correlated according to the off-diagonal term of the above matrix Cx'. Setting directly this *a priori* covariance matrix to such a non-diagonal form would be by far not obvious and it is clear that a different choice would have been made for the off-diagonal term if the inverse problem had been directly set with the two SH and SV velocities as model parameters.

The most objective way to cast the *a priori* covariance matrices of the model parameters is to base them on physical assumptions as was done for example by Nataf et al. (1986) for inverting a global data set of surface-wave dispersion curves in terms of a transversely isotropic model of the upper mantle. Assuming some *a priori* constraints on the upper mantle composition, Nataf et al. (1986) have estimated that the cause of uncertainty in the different parameters was *a priori* linked to the unknowns in temperature, depths of the upper-mantle seismic discontinuities and dynamics of upper mantle materials causing preferred orientation of olivine crystals. Even if some uncertainty exists in the upper mantle composition and in the relationships between temperature and seismic velocities, the estimation of the *a priori* covariance matrix of the model parameters is then based on some physical assumptions and all inverted models based on the same data should be the same, whatever specific choice for the parameterization of the elastic model is made.

4.3 Interpretation of body-wave data

There are two main types of evidence of seismic anisotropy coming from body-wave observations. These are azimuthal variations of refracted-wave velocity and S-wave splitting. A typical example of refracted waves is the observation of azimuthal variations of Pn velocities in the Pacific Ocean (Rait et al., 1969), an observation which corroborates the alignment of the olivine crystal a-axes parallel to the spreading direction of the lithosphere (Hess, 1964; Francis, 1969). Typical examples of S-wave splitting are observed on teleseismic SKS waves traversing the whole mantle (Vinnik et al., 1984), direct-path shear waves for crustal events (Crampin et al., 1980) and VSP (Vertical Seismic Profiling) records in oil seismic exploration (Crampin, 1984a, 1985). In addition to the above pieces of evidence, there is a third type of body-wave observation strongly supporting the possibility of seismic anisotropy in the continental lithosphere: the spatial variations of teleseismic P-wave residuals (see Chapter 5).

4.3.1 Refraction data

Refraction data have been broadly used to infer the seismic velocity structure of the deep parts of the continental lithosphere during the last three decades. In most cases, rectilinear profiles are shot so that no indication of azimuthal variations of velocity is available. A noticeable exception is the study of the continental lithosphere in south Germany (Bamford, 1977). If only one line of direct and inverse refraction profile is available, the study of anisotropy can only be very indirect if one can assume maximum acceptable bounds on seismic velocities according to petrological hypotheses (see the next section). Distinguishing between the S_H and S_V velocities of the refracted shear waves as attempted by Hirn (1977) is difficult due to the poor signal-to-noise ratio of these waves.

The most powerful way of studying seismic anisotropy from refraction data is in fact to measure directly the azimuthal variations of velocity. There are two possible causes of azimuthal variations of refracted-wave velocity: the shape of the horizon on which refraction takes place and the anisotropy of the underlying layer. In case no other lateral variation is present in the overlying layer such observations are not ambiguous. A dipping interface will produce a slow apparent velocity in the dipping direction and a fast velocity in the opposite direction while anisotropy producing a fast velocity in one direction will always produce a fast velocity in the opposite direction, according to the even pattern in the azimuthal velocity variation of Backus (1965). The traditional manner of getting rid of the dipping interface effect in refraction experiments is just to perform direct and inverse profiles, so that the intrinsic velocities of waves propagating just beneath the interface can be measured. For a small dipping angle, the actual seismic velocity V beneath the refraction horizon is given by

$$V^{-1} = (V^{-1}_{down} + V^{-1}_{up}) / 2,$$

where V_{down} is the slow apparent velocity of the refracted wave propagating in the downgoing direction, and V_{up} is the fast apparent velocity in the opposite direction of propagation.

If no constraint is put on the anisotropic velocity model (e.g. hexagonal symmetry with horizontal axis of symmetry) more than two profiles at right angle are necessary to retrieve the anisotropic parameters. Backus (1965) has shown that five lines of reverse refraction profiles are necessary to retrieve the five combinations of elastic coefficients which describes the azimuthal variation of velocity to first order in anisotropy (see the

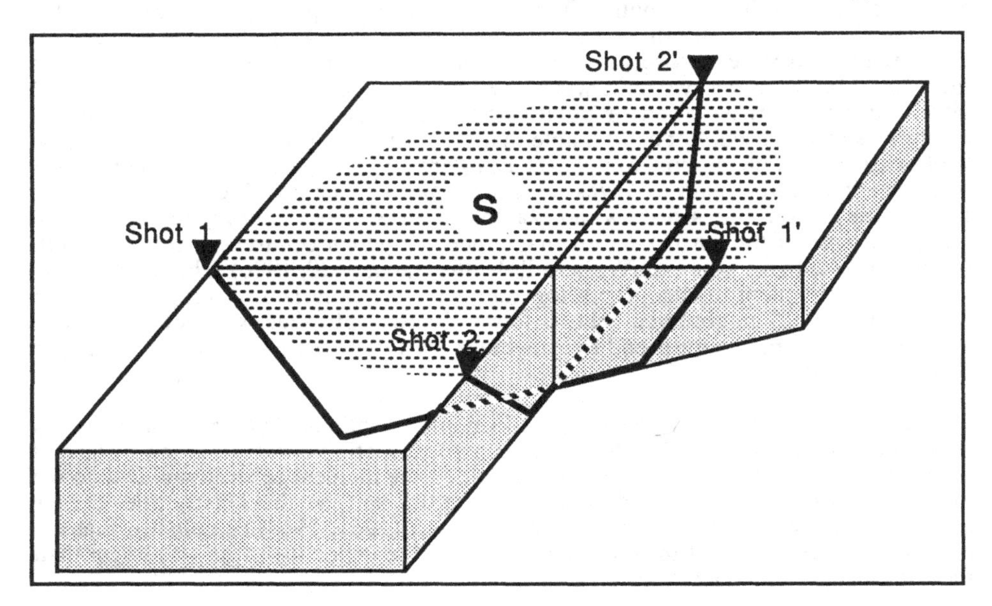

Figure 4-4 Simplified sketch of a refraction experiment designed to study the anisotropic velocity beneath a dipping interface.

formula given in section 2.3) .

The problem of retrieving the anisotropic part of the velocity variations is more complicated in the case of non planar interface and/or lateral heterogeneity of velocities. Two crisscrossing orthogonal refraction profiles, if properly reversed, yield for example two intrinsic average velocities beneath the discontinuity; but they are not sampling exactly the same region, so that inferences on anisotropy of the underlying material can be traded off with the lateral variations of velocities in the shaded area S shown in Fig. 4-4. In the case of a homogeneous upper layer, Backus (1965) has shown that the curvature of the refraction horizon could produce an azimuthal variation of the apparent velocity with an odd angular pattern such as for anisotropy. Only ideas about the maximum velocity variations and the maximum curvature of the discontinuity which are physically acceptable in the investigated area, may allow us to discriminate between the different possible sources of apparent-velocity variations in refraction data. With the sub-Moho seismic velocities in an oceanic environment where the crust is only 6-km thick, this should not be a problem; but on the continent where the crustal thickness is larger and may exhibit strong lateral variations, this might be more problematic.

4.3.2 Shear-wave splitting

The second type of direct observation of the body wave behavior which is a fairly certain indicator of anisotropy is the splitting - or birefringence - of shear waves. We saw in Chapter 2 how a shear wave with rectilinear polarization could be split into two phases with rectilinear polarization at the right angle from each other while entering an anisotropic region. If no discontinuity produces additional splitting in the medium, the two split shear waves will exhibit a travel time difference growing along the ray path, so that it will be possible to observe their different arrival times. In order to measure this time delay between the two split waves, several signal processing techniques based either on simple observation of polarization diagrams or on different correlation techniques can be used. Let us first note that the split shear waves can be observed on the horizontal particle motion. Simple clear evidence of shear-wave splitting comes, for example, from direct-path S waves for earthquake hypocenters located in the crust beneath the station. Fig. 4-5 shows an example of shear-wave splitting observed in the epicentral area of the earthquake of magnitude 1.2 which occurred at a 9 km depth during an earthquake swarm in Western Bohemia (January 20, 1986). The figure shows the three-component seismograms together with the polarization diagrams in the horizontal plane. The observed delay between the two split shear waves S1 and S2 is about 0.1 s. This may correspond to an average 3% S velocity anisotropy in the upper part of the crust.

The basic assumption commonly made in interpretations of polarization diagrams such as that displayed in Fig. 4-5 is that the particle motion observed at the surface of the Earth is the same as the particle motion along the incident seismic ray. Due to the interaction of the incident seismic wave with the free surface, this is only true for rays arriving at the surface with steep incidence. Nuttli (1961, 1964) established that the S-wave polarization was strongly distorted only if the incidence angle was larger than the critical angle at the free surface:

$i_c = \arcsin(\beta / \alpha)$,

where β and α are the S- and P-wave velocities of the underlying medium, respectively. For a normal Poisson ratio of 0.25, this critical angle is close to 35°. Evans (1984b), Booth and Crampin (1985) and Liu and Crampin (1990) have examined in greater detail how the free surface and the internal interfaces of an isotropic structure could affect the

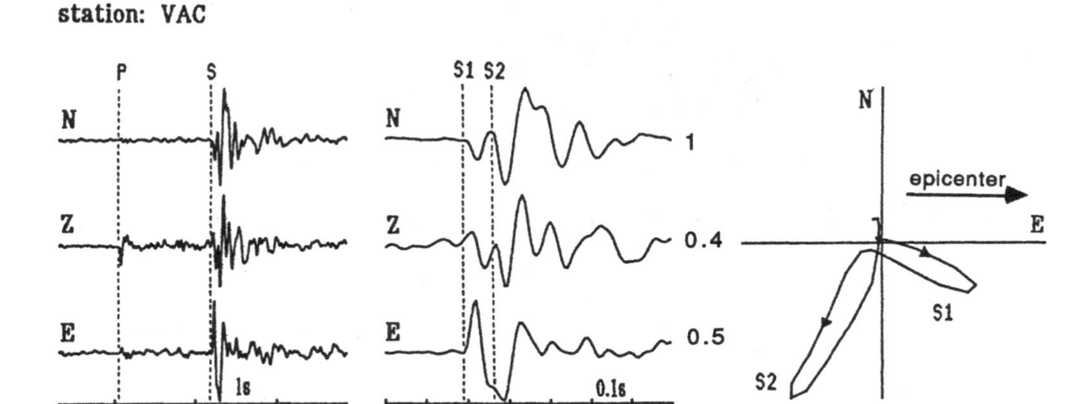

Figure 4-5 Example of direct-path shear-wave particle motion observed at station Vackov located at 6 km from the epicenter of a magnitude 1.2 earthquake, with a focal depth of 9 km, in Western Bohemia (January 20, 1986; after Vavrycuk, 1991). The three-component records with positive motion northward (N), upward (Z), and eastward (E) are shown for two different time scales. Numbers on the right of the records indicate the relative amplitudes. The polarization diagram on the right shows the ground particle velocity in the horizontal plane, with two split shear waves S_1 and S_2.

polarization of shear waves in the case of a curved wave front. Their conclusion is that different windows of the incidence angle can be defined on each interface, the distortion of the S-wave polarization increasing from the innermost window to the external one. Within the innermost window, all the transmission coefficients of shear waves are real and very little distortion occurs. In other windows, the transmitted polarization is more and more perturbed by converted inhomogeneous waves.

The basic difficulty in interpreting observations such as those displayed in Fig. 4-5

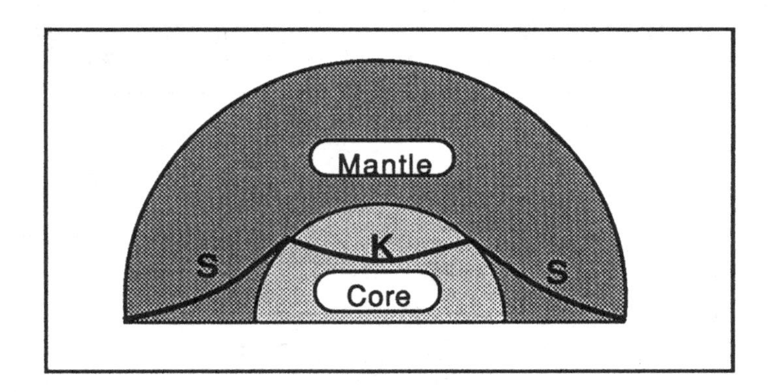

Figure 4-6 SKS wave results from a P to SV conversion at the core-mantle boundary.

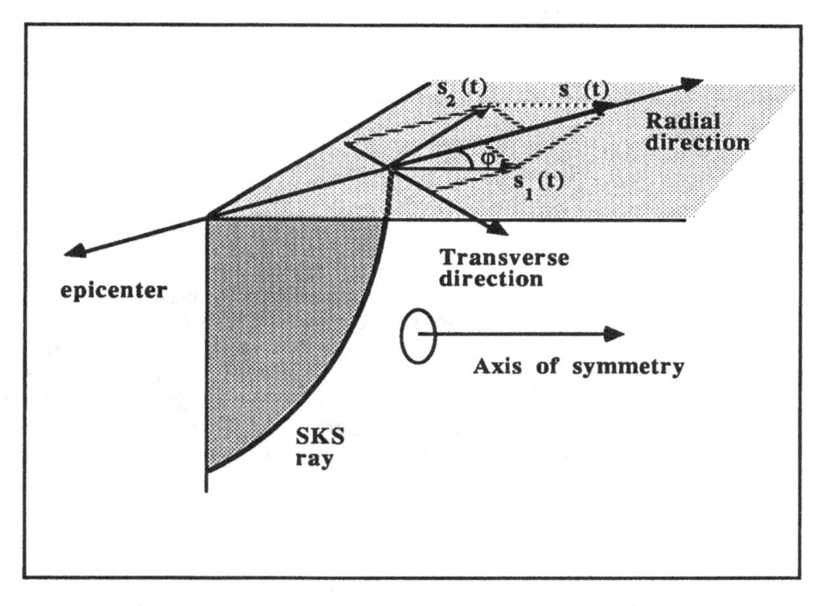

Figure 4-7 Conventions used to derive the relations giving the radial and transverse projections of the two orthogonal S waves $s_1(t)$ and $s_2(t)$ issued from an incident SKS wave s(t). The incident SKS wave s(t) is polarized in the vertical plane and is split into two S waves while travelling through an anisotropic lithosphere represented by a material with hexagonal symmetry and a horizontal axis of symmetry.

is to know if the cause of the apparent shear-wave splitting is due to anisotropy or to wave conversion at a planar interface. Early shear wave arrival could be, for example, an indication of an S_V to P conversion at some distance from the surface. Late shear wave arrivals could also indicate interactions between the curved wavefront of an incident S wave with the free surface (Booth and Crampin, 1985). Surface waves can also be generated at the proximity of the station by the incident S wave interacting with the heterogeneous upper layers or with the topography. We should thus be very cautious when interpreting shear-wave splitting observations by processing several events at different azimuths, if possible by controlling the polarization of the shear wave at the source with the focal mechanism taken into account and, finally, by controlling that the directions of splitting are coherent in the investigated area by analysing the three-component records on several stations for all events and paths.

At teleseismic distances, a very clear example of split shear waves arriving at nearly vertical incidence at the station is provided by SKS waves. SKS travels as a P wave within the liquid core of the Earth. It results from a P to S conversion at the core-mantle boundary (Fig. 4-6). If the layer D" of the lowermost mantle is isotropic, upcoming SKS waves should thus be polarized in the vertical plane in D". If the whole Earth were isotropic, SKS should appear as a pure S_V phase all its way toward the surface. As the SKS ray is nearly vertical beneath the station, it should be observed as a radially polarized phase in the horizontal plane (i.e. tangential to the great circle joining the station to the epicenter). If, on the other hand, there is an anisotropic region somewhere along the upcoming SKS ray path, S-wave splitting will occur yielding two SKS waves polarized at a right angle from each other. Finally, if the symmetry axes of the anisotropic region causing the splitting of SKS have a fixed orientation in the space over a significant portion of the upcoming ray path, for example if the anisotropic parameters are homogeneous in the anisotropic region causing the splitting, the two split SKS waves can arrive at the station with a noticeable time difference, which can be interpreted in terms of preferred mineral orientations in the upper mantle.

Vinnik et al. (1984) and Silver and Chan (1988) noticed that the transverse component of SKS phases, perpendicular to the vertical plane containing the ray path, were quite frequently observed on the horizontal records. In an isotropic symmetrically spherical Earth, no such transverse component is possible. These authors proposed simple models with anisotropy confined to the lithosphere to explain their observations. Silver and Chan (1988) considered for example a homogeneous anisotropy of the lithosphere with hexagonal symmetry and a horizontal axis of symmetry. As we have seen in section 2.3, such a model predicts two principal direction of polarizations for a vertically propagating S wave: one is parallel to the symmetry axis, the other one is perpendicular to it. Assuming that, in a spherical isotropic Earth, the radially polarized SKS signal would be s(t), and assuming the ray is vertical beneath the station, the projection of the ground motion on these two principal directions would be:

$$s_1(t)= s(t) \cos\varphi,$$

and

$$s_2(t) = s(t-\delta t)\sin\varphi,$$

if φ is the azimuth of the fast S-wave polarization with respect to the radial direction which is opposite to the epicenter in the horizontal plane (Fig. 4-7). As the horizontal components of teleseismic waves observed at seismic stations are classically rotated in the horizontal plane so that the seismograms can be analysed along the radial and transverse components it is interesting to compute the radial and transverse components of the split SKS waves. Projecting the two split signals $s_1(t)$ and $s_2(t)$, one gets for the radial component:

$$R(t) = s(t)\cos^2\varphi + s(t-\delta t)\sin^2\varphi,$$

and

$$T(t) = [(s(t) - s(t-\delta t))/2]\sin 2\varphi,$$

for the transverse component.

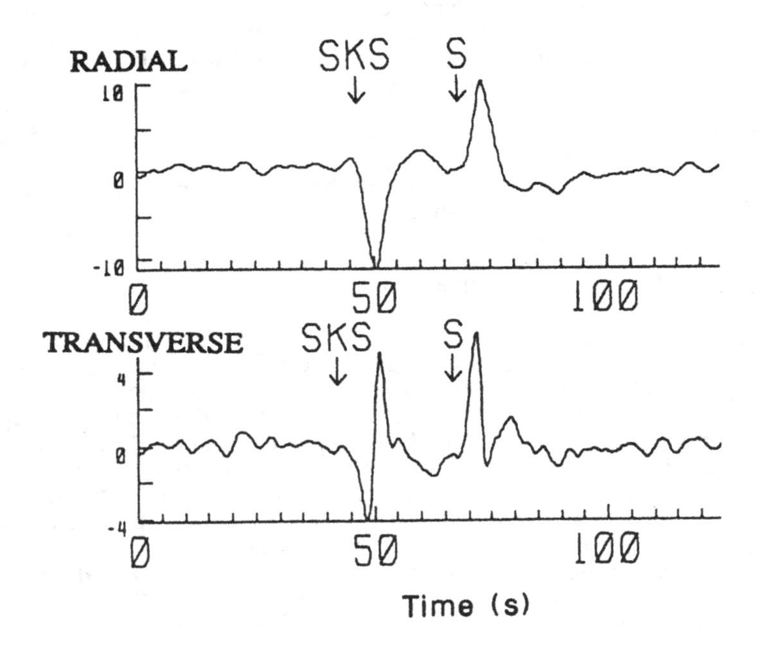

Figure 4-8 Example of split SKS phases observed at the station RSON in Canada for a deep event at 89.1° epicentral distance. The transverse component of SKS should not appear if the Earth were representable by an isotropic symmetrically spherical model. The abnormal transverse component of SKS appears as the time derivative of the radial component (from Silver and Chan, 1988). (Reprinted by permission from *Nature*, Copyright © 1991 Macmillan Magazines Ldt).

For a weak anisotropy occurring within the lithosphere, δt tends to zero and one can expect that the waveform observed on the transverse component T(t) should appear as the first derivative of the waveform observed on the radial component R(t). Actually this is what has been quite systematically observed by Silver and Chan (1988): out of the 44 records of SKS waves analysed by these authors, 28 have a visually observable transverse component of SKS with time differences δt between the two split waves up to 1.75 s. A particularly clear observation of a split SKS wave is shown in Fig. 4-8.The simple model presented to explain the observation shown in Fig. 4-8 may not work in other situations. First of all, the cause for anisotropy is entirely located within the lithosphere where the ray is nearly vertical. Deeper causes of anisotropy were ruled out by Silver and Chan (1988) in their study. Second, the anisotropic properties have been modelled assuming a hexagonal symmetry with a horizontal axis of symmetry. If the cause for the observed splitting is due to preferred orientation of olivine crystals, an orthorhombic symmetry with not necessarily horizontal and vertical symmetry axes would have been a more suitable model closer to the reality. Among these different remarks, however, it is likely that the most important one is that related to the uniformity of the symmetry axis orientation. If no such uniformity exists, more than two split phases would be observed with small time delay between them, making the interpretation difficult . Examples of complicated SKS splitting have been analysed by various authors, e.g., by Vinnik et al. (1989a) and Ansel and Nataf (1989).

Heterogeneity is often opposed to anisotropy as an alternative explanation of seismic observations. Heterogeneity of the anisotropic structure may of course also complicate the situation. The paradox is that a too large heterogeneity in the orientation of the symmetry axes of the anisotropic material along a ray path may destroy the overall anisotropy and make isotropic modelling of the observed seismic wave more straightforward!

4.4 Long-range refraction profiles

Long-range seismic refraction profiles with controlled sources and lengths between about 200 km and 3500 km provide important information on structure and properties of the continental upper mantle (see e.g. Fuchs and Vinnik, 1982, for a review). The most reliable data are provided by reversed and overlapping profiles with determined crustal structure and with average distance between the neighboring stations of less than 10-20 km. The observations are interpreted in terms of a velocity-depth distribution in the upper mantle. A comparison of the observed and theoretically computed phases is often used for this purpose. The phase correlation is a relatively simple and reliable procedure for first arrivals, but could be more difficult for later arrivals.

There are two basic philosophies in interpreting the data of long refraction profiles (Masse, 1987). One philosophy, followed by a number of European seismologists (e.g. Mueller and Ansorge, 1986), interprets a disappearance of a refraction phase in a profile as an evidence for a low-velocity layer. Such interpretations lead to construction of short discontinuous refraction lines and to a complex lamellar structure of the upper mantle consisting of layers with alternating high and low P velocities through the entire depth range. The other philosophy (e.g. Masse, 1987) interprets the disappearance of a refraction phase as a simple data gap which is not necessarily caused by a low-velocity layer. Such interpretation yields relatively simple velocity-depth models, although a part of the information seems to be lost.

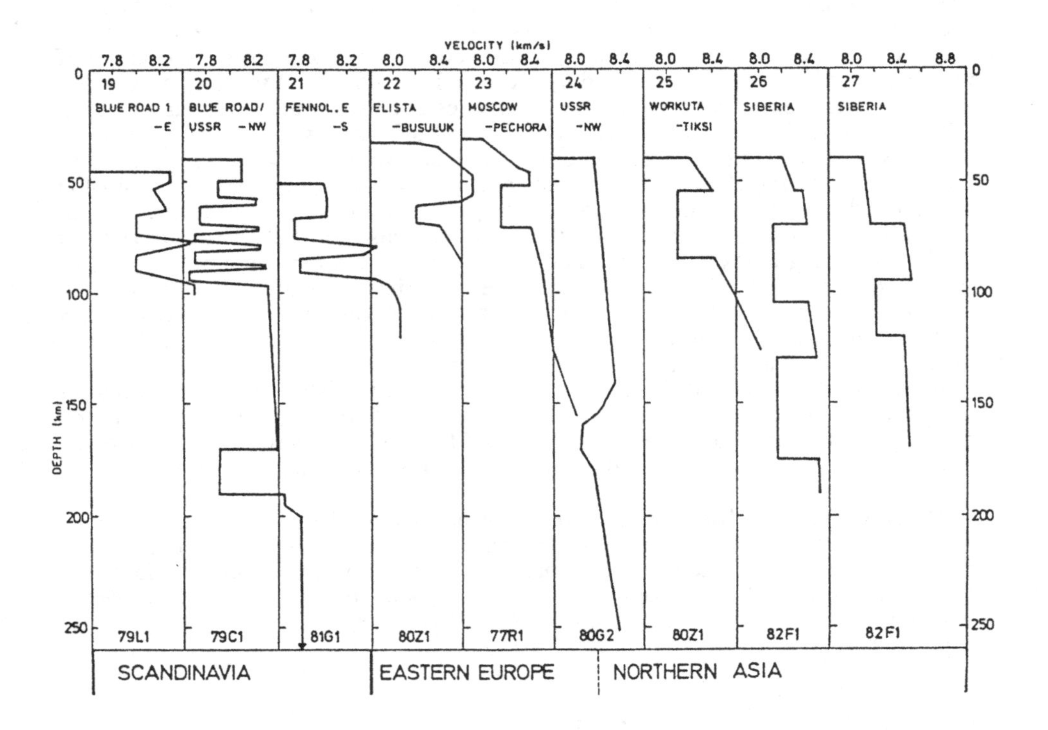

Figure 4-9 One-dimensional velocity-depth models of the uppermost mantle in Eurasia (Fuchs et al., 1987). (AGU, copyright © 1987).

Results of observations on long refraction profiles are usually presented in the form of a one-dimensional velocity-depth distribution in the upper mantle. Prodehl (1984) and Fuchs et al. (1987) summarized such distributions for several regions of Eurasia (Fig. 4-9). It is a great simplification to characterize regions extending sometimes several thousand of kilometers by a single velocity-depth profile, because many observations show that not only the crust but also the upper mantle of continents are laterally heterogeneous. Therefore some authors aimed to construct from several one-dimensional models two-dimensional models with lateral changes in the velocity distribution. Fig. 4-10 shows as an example of such models, the Fennolora profile (Guggisberg, 1986), which extends over two thousand kilometers across the Fennoscandian Shield.

Both types of the velocity-depth distribution characterize the uppermost mantle by a sequence of alternating high- and low-velocity layers (Figs 4-9 and 4-10). However, the large differences in interpreted P velocities between neighboring layers, which are

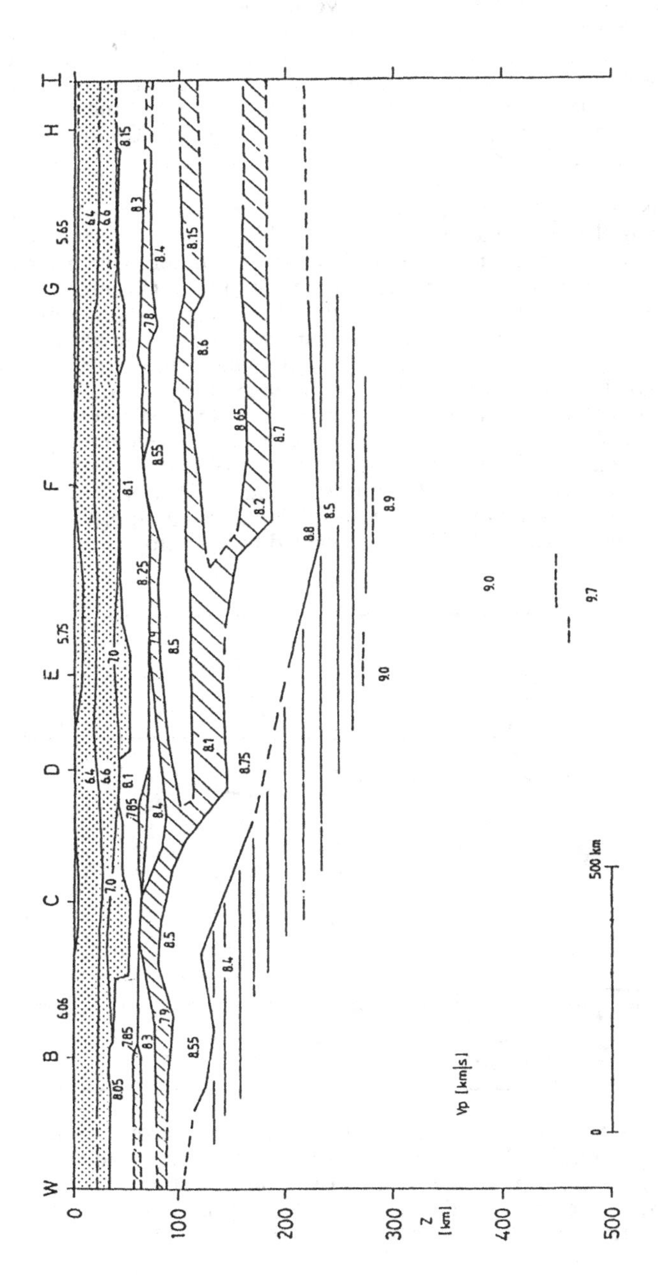

Figure 4-10 Two-dimensional velocity-depth models of the lithosphere beneath the Fennoscandian Shield along the international Fennolora seismic refraction profile (Guggisberg, 1986). The crust is dotted and the low-velocity layers in the upper mantle are hatched.

as large as 0.5 - 0.6 km/s, as well as overall velocity variations between 7.7 and 8.8 km/s, are practically impossible to explain in terms of isotropic petrological models (Babuska and Gueguen, 1989) deduced from the world-wide sampling of mantle xenoliths. Only unrealistic quantities of garnet or spinel could explain the isotropic P velocities as large as 8.8 km/s, and there is probably no mineral which could decrease the isotropic velocities below 8 km/s under upper mantle physical conditions. On the other hand, with preferred orientation of olivine as the major component of mantle rocks and its consequence in anisotropic propagation of seismic waves (see Chapter 5) the velocity variation can be easily explained.

Several observations on long refraction profiles proved the existence of seismic anisotropy in the continental upper mantle. Bamford (1973, 1977) determined an azimuthal variation of Pn velocity amounting to 7-8 % beneath southern Germany and found also a weak anisotropy beneath the western regions of U.S.A. (Bamford et al., 1979). Hirn (1977) explained azimuthal variations of travel times and amplitudes of P waves and possible birefringence of S waves beneath western Europe by an anisotropic layer at depths of about 60-100 km. Vinnik and Egorkin (1980) suggested seismic anisotropy as a possible explanation for substantial differences between the velocity models to depths of about 200 km obtained along two almost perpendicular profiles across the Siberian Platform. While the latitudinal profile showed low-velocity layers at depths of about 70-100 km and 130-170 km, at the meridional profile the authors found at these depths high-velocity layers (Fig. 4-9, columns 26 and 27). For intersecting profiles this seems to be a strong argument in favour of the anisotropy, although the authors admitted that the observations could be also explained by a heterogeneity.

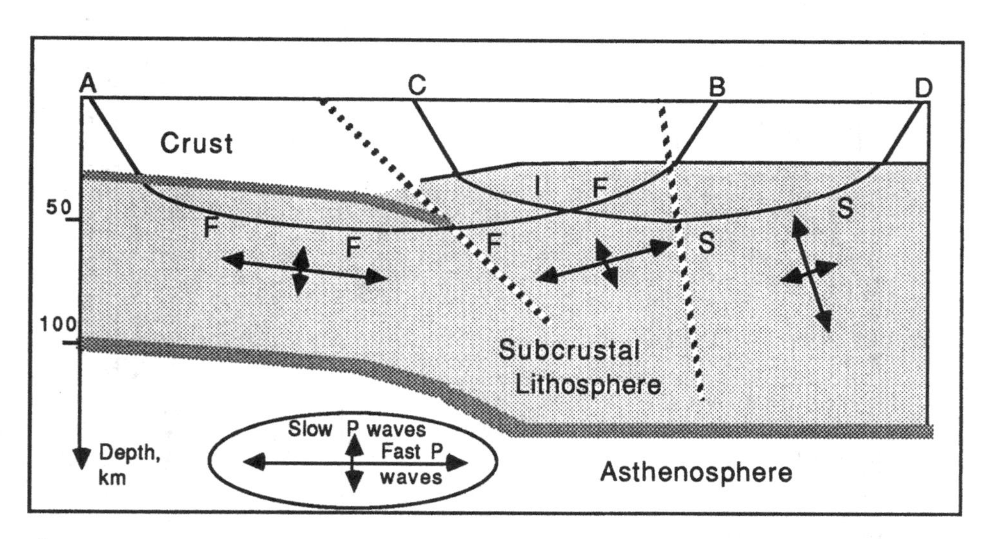

Figure 4-11 Schematic two-dimensional model with dipping anisotropic structures in the deep continental lithosphere and ray paths of refracted seismic waves. F, S and I indicate fast, slow and intermediate P-wave velocities, respectively. The wave propagation over ray AB is abnormally fast while it is abnormally slow over ray CD.

Besides the anisotropy observations on crossing profiles, there are also indirect indications of anisotropy in extremely high P velocities (8.6-8.9 km/s) at depths of about 50-100 km (e.g. Fuchs, 1977; Ansorge et al., 1979). The vertical velocity gradient reaches 0.03 s^{-1} in southern Germany, which exceeds the gradient in a homogeneous geothermally heated medium under self-compression by about two orders of magnitude. In the absence of phase transition with sufficiently strong velocity increase, the high velocities in combination with the extreme velocity gradient provide a strong indirect evidence for the anisotropy in the subcrustal lithosphere of southern Germany (Fuchs et al., 1987). As will be shown in Chapter 5, this has been confirmed by later independent observations.

Which types of anisotropic structures can explain the velocity variations interpreted from data of long refraction profiles? Fuchs et al. (1987) suggested that the high-velocity layers could be produced by a preferred orientation of olivine and the low velocities either by the absence of a preferred orientation or by a decreased concentration of olivine. However, it is very difficult to imagine a mechanism which could produce such different olivine orientations in alternating subhorizontal layers with thicknesses from several kilometers to several tens of kilometers. Therefore we suggest another model, which explains the velocity variations observed along the refraction profiles by a propagation of seismic waves in systems of dipping anisotropic structures in the subcontinental lithosphere.

As we will show in Chapter 5, the continental lithosphere most probably consists of blocks which are several hundred kilometers in dimensions. We assume that the blocks were created in a period of successive accretions of oceanic lithosphere, or, by the accretion of an older continental fragment, which also retains remnants of paleosubductions of anisotropic oceanic lithosphere by which the continents grew in the past. The blocks are separated by deep faults, or sutures, where abrupt changes in the thickness of the crust and the whole lithosphere are often observed.

We thus suggest that the fundamental property of the wave fields of long refraction profiles - the presence of several mantle phases replacing each other as first arrivals with increasing distance - is not caused by a sequence of low- and high-velocity layers in the mantle, but by lateral changes of orientations of large-scale anisotropic structures at the boundaries of the lithospheric blocks (Fig.4-11). We could assume that the anisotropy pattern and its orientation in the subcrustal lithosphere do not change significantly within individual blocks, if each of them corresponds to a piece of subducted oceanic lithosphere with frozen-in anisotropy (see section 5.5.3). As the orientation of anisotropic structures does change at the boundaries of the lithospheric blocks which represent products of different phases of continental accretion (Babuska and Plomerova, 1989), we can assume that, e.g., a disappearance of a refraction phase in the data of seismic profile need not always be caused by a low-velocity layer, but rather by a lateral change in the orientation of anisotropy. At different epicentral distances from the source we could thus observe differences in arrival times as compared to an "average" model which depends on orientations of anisotropic structures along the wave paths (Fig. 4-11).

More than a decade ago, Oliver (1980) wrote about "the inadequacy of spherical or flat-layered models" that portray the continental basement and our "level of understanding has reached the point where refinement of such models may be more misleading than informative". Recent seismological observations such as SKS splitting, which proved the existence of large-scale anisotropy in the deep continental lithosphere, support his ideas. Providing that future observations on tightly spaced detection networks containing 3-component stations for a detection of S waves enable us to characterize at least approximately lateral changes in orientation and symmetry of the anisotropic structures, then 3-D models of the continental lithosphere composed of autonomous anisotropic blocks could become a more realistic alternative to the simple models of high- and low-velocity layers.

4.5 Interpretation of surface-wave data

The fact that the upper mantle anisotropy could be observable on surface wave dispersion curves was first mentioned in the early sixties (Anderson and Harkrider, 1962, Anderson, 1966). The first surface-wave measurements showing clearly that Love- and Rayleigh-wave dispersion curves could not be compatible with a unique isotropic model were obtained by Aki and Kaminuma (1963) for Japan and by McEvilly

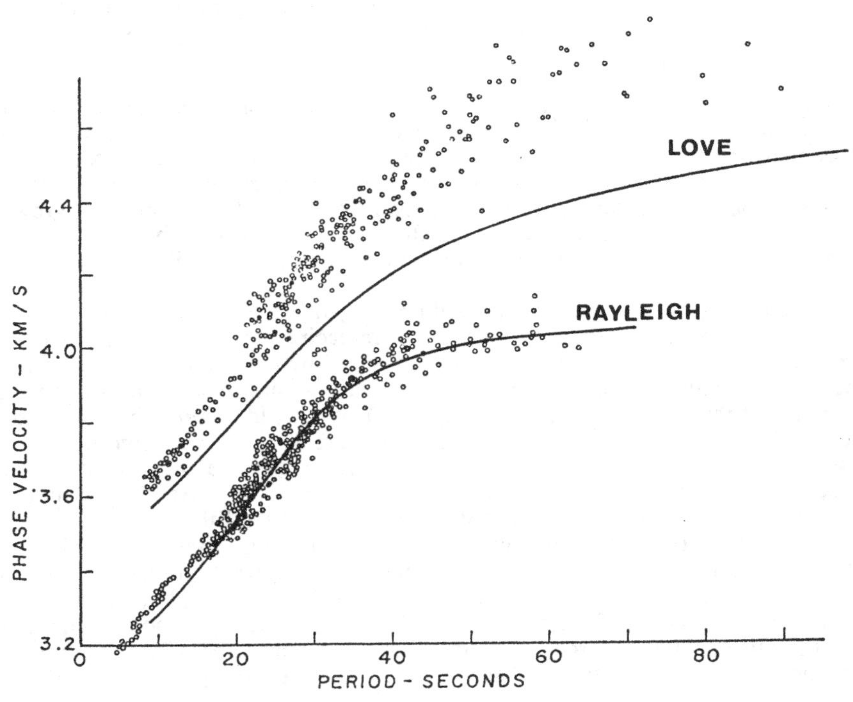

Figure 4-12 Love/Rayleigh-wave discrepancy observed by McEvilly (1964) in central US. The observed phase velocities of the Love-wave fundamental mode are too high, compared with the velocities predicted from a Rayleigh-wave study in the same region. (With permission of the Seismological Society of America, copyright © 1964).

(1964) for the United States. Fig. 4-12 by McEvilly (1964) shows the Love- and Rayleigh-wave dispersion curves measured in central U.S. together with theoretical curves computed for an isotropic model fitting the Rayleigh-wave data. Such a type of observation has been the subject of considerable debate, the question being whether these observations prove the existence of seismic anisotropy at depth or not. As already stated, isotropic models made of thin horizontal layers with alternate high and low rigidities can explain such data as well as an intrinsically anisotropic structure; but this point was not really at issue in the debate. It concerns other possibilities of interpreting the data.

Boore (1969) and James (1971) substantiated the observations of too high velocities of Love waves by an interference effect between the fundamental mode and the first overtone. Indeed, in an oceanic environment, the group velocities of both the fundamental mode and the first higher mode of Love waves - around 4.4 km/s within a broad period range - strongly suggest that interference effects can affect the dispersion measurements due to the proximity of the two modes in the time-frequency domain. Forsyth (1975) was the first to oppose this argument by computing the excitation of both modes at the source for realistic source parameters and an appropriate elastic structure in the source region. Computing the excitation level of the fundamental mode and first overtone of Love waves, he found that the amplitude of the first overtone was far too small to explain how interferences with the fundamental mode could significantly alter the dispersion measurements.

An answer to the Boore (1969) and James (1971) suggestion that the too high Love-wave velocity observation was a pure artifact coming from an interference effect with higher modes came from the application of stacking techniques to very large aperture arrays of long-period seismic stations. Such stacking techniques were developed in the mid-seventies to perform dispersion measurements of surface-wave overtones (Nolet, 1975; Cara, 1978). In order to get rid of the interference effects, due to the too close proximity of group arrival-time curves in the time-frequency domain, the idea was to analyse the seismic wavefield in terms of its wavenumber content. Since, for a given frequency, the different modes of either the Love or Rayleigh waves have different phase velocities, and thus different wavenumbers, the wavenumber analysis techniques can handle the problem of interference between the different surface-wave modes and give us the individual dispersion curves of each mode.

The first application of such a technique to Love waves was made by Cara et al. (1980). A clear Love/Rayleigh-wave discrepancy was observed in western Europe for both the fundamental mode and the first overtone, the phase velocity of the Love waves being too fast by about 0.2 km/s compared with the values predicted from elastic models built up from the inversion of Rayleigh-wave data in the same region.

Another application of such techniques to the measurement of Love- and Rayleigh-wave dispersion curves along epicenter-station paths was performed by Leveque and Cara (1983) in the Pacific. Dispersion curves up to the third overtones were obtained over paths about 10 000 km long crossing the whole Pacific (Fig. 4-13). Both wavenumber analysis techniques and estimation of the source phase are necessary to extract the dispersion curves of several modes from a set of seismograms (Cara, 1978). The respective data set is shown in Fig. 4-14 together with the curves predicted for the isotropic model P7 of Cara (1979) which fits the Rayleigh-wave dispersion curves on similar paths. A clear Love/Rayleigh-wave discrepancy is observed both for the

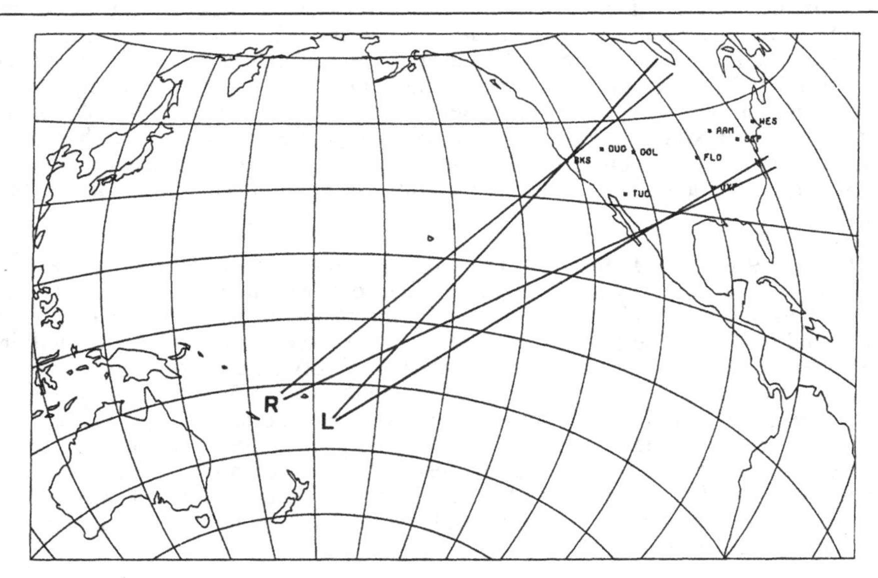

Figure 4-13 Paths used by Cara (1979) and Leveque and Cara (1983) to measure overtone dispersion curves of surface waves in the Pacific. Rayleigh and Love data were measured for events located in the areas marked R and L, respectively. The WWSSN stations used for Love waves are indicated on the map. (Reprinted from Leveque and Cara, 1985, with permission of the Royal Astronomical Society, copyright © 1985).

fundamental mode at all periods and for the overtones at the shortest periods: the observed Love-wave dispersion curves are too fast compared with the curves predicted from Rayleigh-wave model P7. Because the stacking method used in this study is designed to eliminate the interferences between the different mode branches, artifacts linked to interferences between modes can be ruled out in this data set. The region related to these dispersion curves is a 10000-km long area of oceanic lithosphere with an average age of 70 Ma. In this study, the only strong lateral heterogeneities are within the terminal part of the paths to the seismic stations, in the continental part of the United States where the seismic array is located. It is very unlikely, however, that the artifact linked to the lateral heterogeneities within the array of stations can be the cause of the observed anomaly in the observed Love-wave velocity since, first, the dispersion data are corrected for the effect of propagation in the continental part, and second, this continental part, even if poorly taken into account in the correction, affects less than 20% of the total ray paths.

If the dispersion measurements are made along a single azimuth as was done by Leveque and Cara (1983) in their study of the Pacific, there is of course no way of getting information on the azimuthal variation of the surface-wave velocities. If there are independent reasons to think that the azimuthal variation of velocities is negligible, transversely isotropic models with a vertical axis of symmetry are the best suited for interpreting the data. When the dispersion data are interpreted in this way, it is indeed preferable to use transversely isotropic models with five elastic coefficients varying with depth than to use two independent isotropic models to interpret the Love- and Rayleigh -wave dispersion curves. Such a procedure, which was commonly applied in the past, can lead to erroneous conclusions (Kirkwood, 1978). Without invoking basic questions on the physical significance of such an independent inversion of two sets of data related to the same geographical area, let us just point out that the depth-resolution curves associated with the two sets of independent Love- and Rayleigh-wave dispersion curves are not in general identical. Therefore, we cannot draw any inferences upon anisotropy only from the discrepancies between the elastic models derived from two data sets that represent the Earth with different depth resolutions. Simultaneous inversion of the two data sets within the framework of correct anisotropic modeling and with a careful

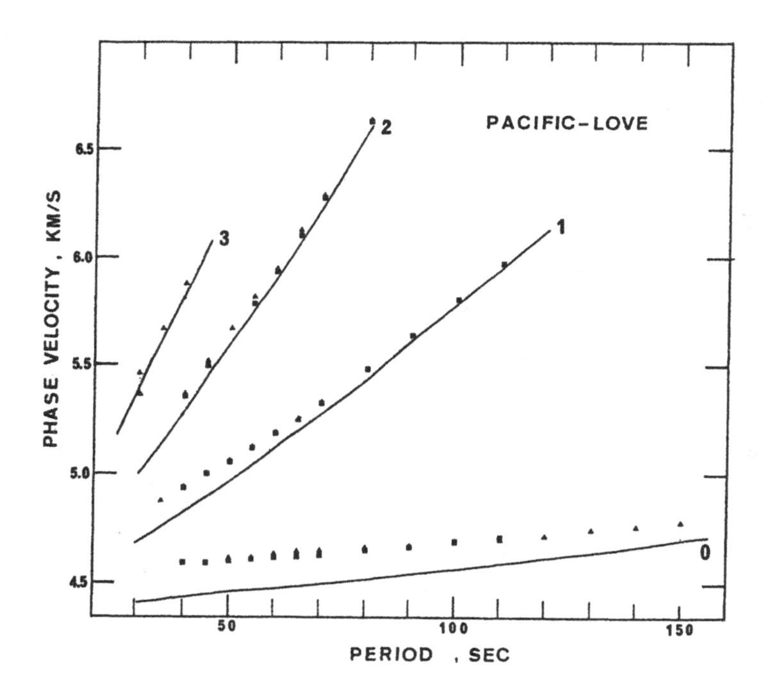

Figure 4-14 Love-wave phase velocities in the Pacific. Symbols show the observed phase velocities for two different events (mode 0 for the fundamental mode and 1 to 3 for the first three overtones). The theoretical dispersion curves computed for an isotropic model fitting the Rayleigh-wave data up to the third overtone, in the same region are shown for comparison by the curves (after Cara and Leveque, 1987, 1988). The incompatibility between the observed and theoretical dispersion curves can be explained by upper-mantle anisotropy. (AGU, copyright © 1988).

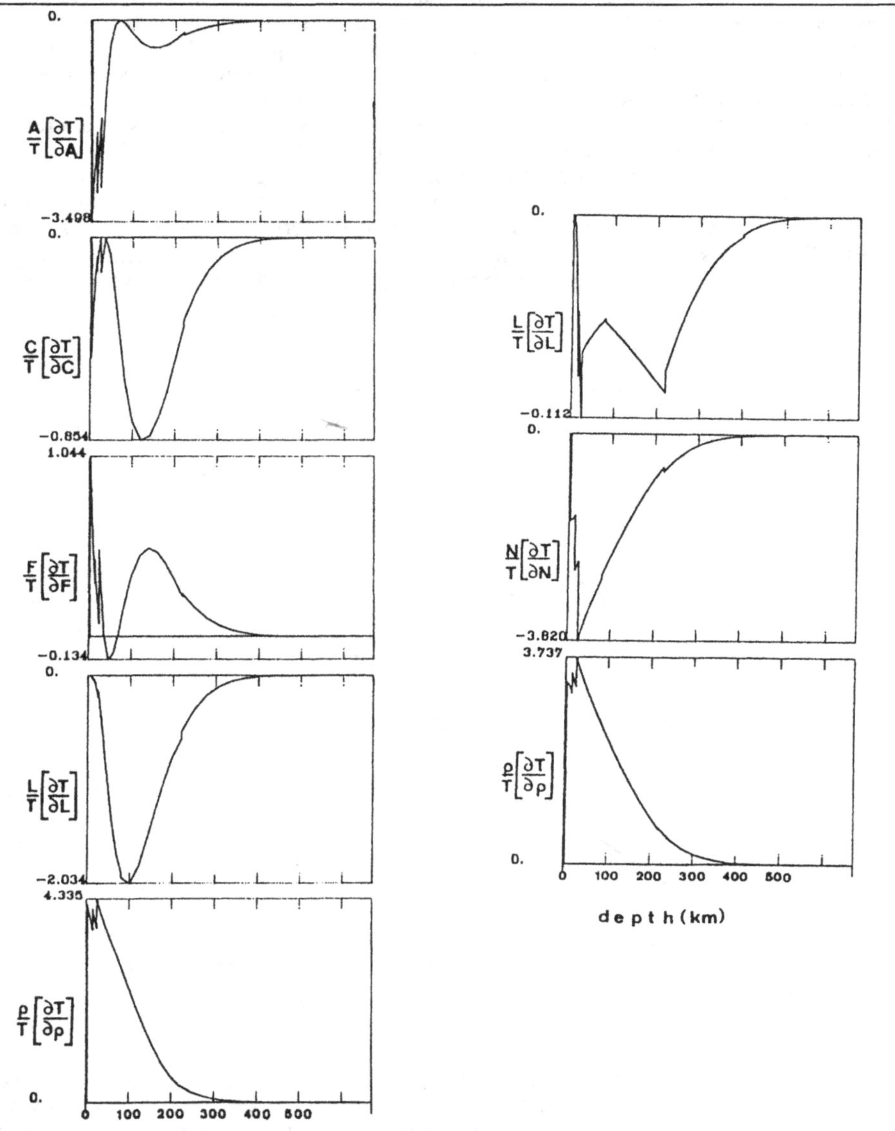

Figure 4-15 Partial derivatives of the eigenperiod of vibrations of the fundamental mode of spheroidal mode at 82 s (Rayleigh waves - left column) and toroidal mode at 73 s (Love waves - right column) with respect to the five elastic coefficients of a transversely isotropic Earth's model: A, C, F, L, N. The curves are normalized to 1-km layer thicknesses (after Montagner and Nataf, 1986). (AGU, copyright © 1986).

examination of the resolution curves of the anisotropic parameters is a much safer way to interpret the data, even if it is a more sophisticated procedure to implement on the computer.

In the case of global observations of dispersion curves, sampling the Earth's surface along great circles with various azimuths and poles, the transversely-isotropic modeling is well suited for interpreting the data. Five independent elastic coefficients varying with depth are needed to describe a spherical transversely isotropic model. Starting an inversion process within such a transversely isotropic framework, we have to compute first the partial derivatives of the eigenperiods of vibrations of the whole Earth with respect to these five elastic coefficients and the density. The equations to compute the partial derivatives of the eigenperiods of vibrations of toroidal and spheroidal modes are given by Takeuchi and Saito (1972). For progressive surface waves, it is more convenient to convert these eigenperiod partial derivatives in terms of phase-velocity partial derivatives. Toroidal modes correspond to Love waves and spheroidal modes to Rayleigh waves. Examples of such partial derivatives of the fundamental modes of Love and Rayleigh waves are depicted in Fig. 4-15.

The parameterization chosen in this figure is in agreement with Takeuchi and Saito (1972), with two seismic velocities and three dimensionless anisotropic parameters ξ, φ and η (as in section 4.2). As is usual in surface-wave studies, the compressional velocity is a much less sensitive parameter than the shear wave velocity. In order to make possible the comparison between these different curves, they have been normalized so that each curve gives the phase-velocity perturbation when perturbing an elastic coefficient over a one-kilometer thick layer. Note that varying the ratio of S_H to S_V velocities by 1% means that ξ is changed by 2%.

Such partial derivative curves can be used for interpreting phase velocities observed between two stations or between one station and the epicenter of an earthquake if no azimuthal variation of the phase velocity is allowed for in the inversion. Such a technique was applied, for example, by Regan and Anderson (1984) to interpret a large set of Love and Rayleigh dispersion curves related to the Pacific. In this study, lateral variation of velocity was allowed for but no local azimuthal variation of velocity was considered.

On the global scale, one of the first attempts to invert a large set of data in the framework of such a transversely isotropic model was made by Dziewonski and Anderson (1981). The laterally homogeneous spherical model they derived is known as the Preliminary Reference Earth Model, or PREM, which is shown in Fig. 4-16. This model is made of five elastic coefficients plus the density varying with depth. Small anisotropy is present in the upper part of the model (see Fig. 4-16), a few per cent velocity anisotropy being present down to the 220 km-depth discontinuity of the model. Model PREM was broadly used as a starting model in global and regional studies during the 1980's.

Resolution curves for the six physical parameters of PREM which were estimated by Nataf et al. (1986) are worth commenting on as they provide an objective way of assessing the validity of the anisotropic parameters in this model. Nataf et al. (1986) estimated the PREM resolution kernels from a subset of 120 normal mode data used to build up the model. Since these data are the most sensitive to the anisotropic parameters, the resolution kernels they have computed are pertinent to answering the question of the

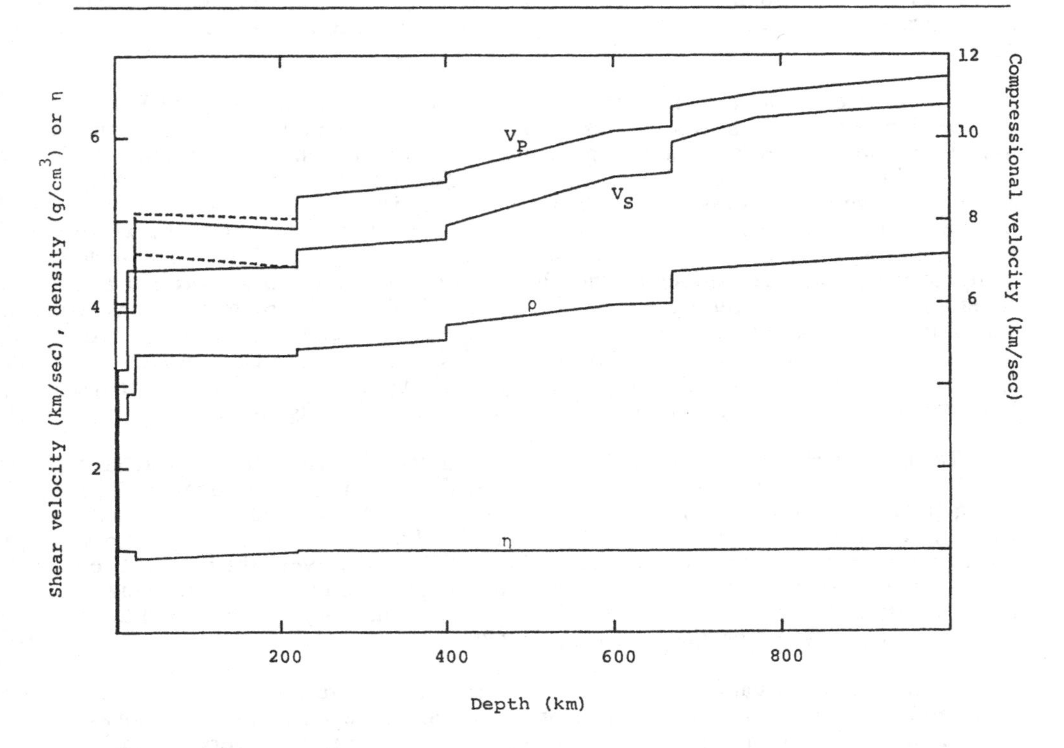

Figure 4-16 Transversely isotropic global model PREM of Dziewonski and Anderson (1981). The model is described by the five parameters α_V, β_V, ξ,φ and η. ξ and φ are not shown in this figure. Note that no anisotropy is required by the data below 220 km. (With permission of Elsevier Science Publisher, copyright © 1981).

reliability of anisotropy in PREM. The kernels were estimated with the use of the formalism of Tarantola and Valette (1982) under the assumption that there were no *a priori* correlations between the six parameters and no *a priori* correlations between two successive data points along the depth coordinate either. A correlation length of 1 km along the depth coordinate was set and *a priori* standard deviations of the parameters were: 0.1 g/cm³ for the density ρ; 0.1 km/s for P_V and S_V velocities, α_V and β_V; 0.1 for ξ, φ and η.

The resulting resolution curves are shown in Fig. 4-17. Very poor resolution is obtained for the compressional velocity α_V as is usual when we use only normal mode data. Actually, the P velocities of the PREM model were constrained by travel time data which are not taken into account in the computation of these resolution curves. Good resolution is obtained for vertically polarized shear velocity β_V in the depth range

300-400 km. At a shallower depth the shear velocity is traded off with the anisotropic parameters and the density, probably due to a lack of short enough periods in the data. Fig. 4-17 shows that the shear-wave anisotropy ξ is one of the best resolved parameters in PREM. It exhibits very little trade-off with the other parameters, except with the density ρ. On the other hand, the two other anisotropic parameters φ and η are strongly traded-off and thus do not furnish independent information. Finally, density ρ is well resolved but it exhibits some trade-off with the anisotropic parameter ξ and the shear wave velocity β_V. The fact that density is traded-off with anisotropy is an interesting by-product of this resolution analysis which shows that care must be taken when we derive density models from normal mode data. This point was noticed by Forsyth (1975) and Cara et al. (1984) while they were trying to get better constraints on the density versus depth variations from surface-wave data.

The resolution curves presented in Fig. 4-17 show that shear velocity anisotropy can indeed be resolved from the PREM data for the upper part of the mantle. The fact that the departure from isotropy is small is one of the best constraints for setting upper bounds on the overall anisotropy in the Earth on the global scale. Of course, this does not give any information on the bounds to be put on the anisotropic parameters on a regional scale.

An attempt at constructing a global anisotropic model taking into account lateral heterogeneities of the velocities in a transversely isotropic model was made by Nataf et al. (1986). On the global scale, this approach is similar to the approach adopted by Regan and Anderson (1984) on the scale of the Pacific plate. Locally, the Earth is supposed to be transversely isotropic. It is thus described by six independent parameters varying with depth. More precisely, transverse isotropy is assumed within areas of the size of the lateral resolution of the regionalization method - a few thousands of kilometers in this study, which is based on a set of 200 great-circle paths for Love waves and 250 paths for Rayleigh waves. This approach should not pose any problem if, within these few thousand kilometers wide areas, crisscrossing surface-wave rays sample regularly a wide enough range of azimuths. By averaging locally the surface-wave velocities of different azimuths, the regionalization procedure may build up a transversely isotropic velocity model which would be equivalent to averaging locally the actual anisotropic structure by rotating it around the vertical axis.

The regionalization procedure used by Nataf et al. (1986) is twofold. First, according to the division of the surface of the Earth into the seven tectonic provinces proposed by Okal (1977), the global set of phase and group velocity curves was regionalized by a simple least square method. Second, a spherical harmonic expansion of observed surface-wave velocity was performed up to the angular order 6 and the period-dependent coefficients of the expansion were regarded as dispersion data to be inverted like classical dispersion curves. In a second step, the data regionalized either by tectonic provinces or by spherical harmonic expansion were inverted in terms of depth-varying properties. One of the most interesting conclusions of this study is that the lateral variations of shear-wave velocities which were previously obtained from similar data in the framework of an isotropic Earth model by Woodhouse and Dziewonski (1984) are not drastically affected by the introduction of anisotropy in the model down to a depth of about 400 km. At a greater depth, large discrepancy occurs; but one must notice that the lateral variation of shear-wave velocity is then much smaller. Trade-off between anisotropy and lateral heterogeneity does occur but it appears to be small enough so that the gross images previously obtained within the

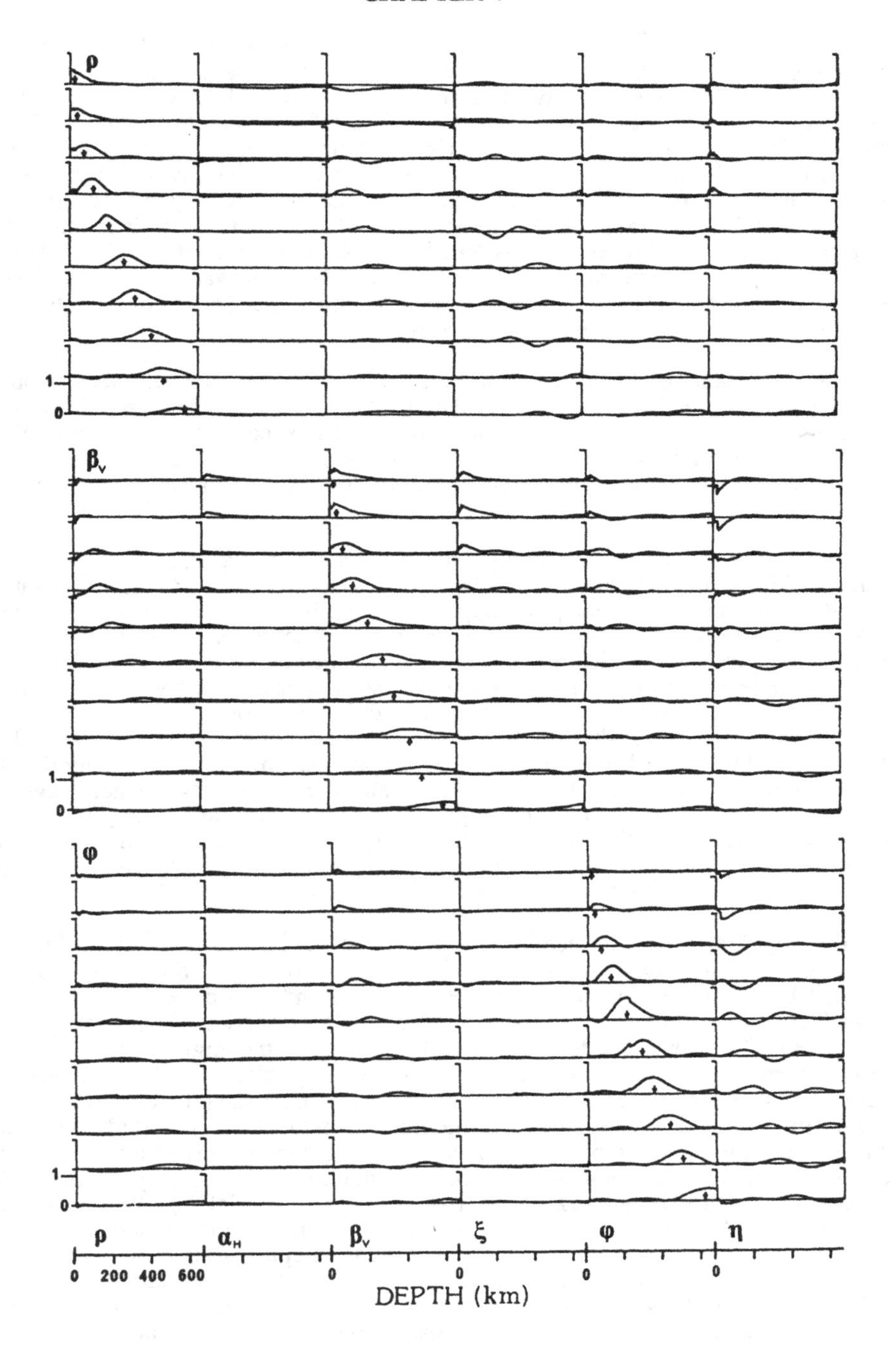
ρ
β_V
φ
1
0
ρ
α_H
β_V
ξ
φ
η
0 200 400 600
0
0
0
0
DEPTH (km)

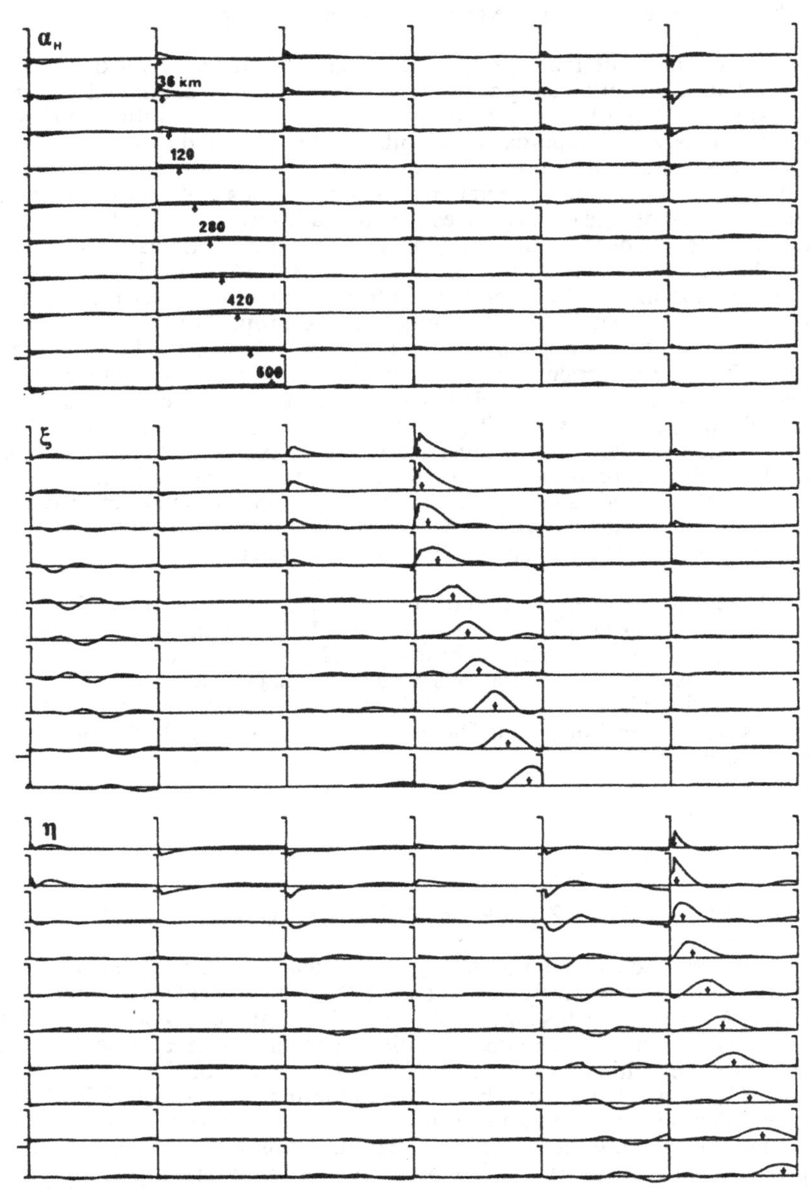

Figure 4-17 Resolution curves of transversely isotropic Earth model PREM estimated by Nataf et al. (1986) from the subset of 120 normal mode data used by Dziewonski and Anderson (1981) to build up the model. Units on the vertical scale are 10^{-5} km^{-1}. Model parameters are normalized to their standard deviations. (AGU, copyright © 1981).

framework of isotropic surface-wave tomography remain valid.

In order to interpret the surface-wave data showing a Love/Rayleigh-wave discrepancy caused by an actual physical situation involving azimuthal variation of velocity, it is possible to follow the mixed approach of hypothesis-testing and inversion mentioned previously. This approach was followed by Cara and Leveque (1988) who assumed that the cause of the observed Love/Rayleigh-wave discrepancy displayed in Fig. 4-14 was due to a preferred orientation of olivine crystals in the upper mantle of the Pacific Plate with the olivine crystal a-axes preferentially oriented in the horizontal plane (see Fig. 6-6). For this purpose the direction of the preferred orientation of the a-axis was one of the tested parameters while the b- and c-axes were randomly oriented around this horizontal direction. The overall anisotropic model exhibits then hexagonal symmetry with a horizontal axis of symmetry. The proportion of oriented olivine crystals is the only free anisotropic parameter in this experiment. The best fit to the data is obtained if the preferred-orientation direction of the olivine a-axes is set parallel to the present-day direction of the absolute plate motion. The result is a well constrained upper mantle model since, eventually, two depth varying parameters - one isotropic, the other anisotropic - were considered by the inversion process. This example illustrates the fact that, even if no information is available on the azimuthal variation of velocities, it may be worthwhile testing models that predict such an azimuthal variation instead of using transversely isotropic models with vertical axis of symmetry which may have no physical significance in a single-azimuth dispersion measurement.

Further steps in the interpretation of surface-wave data have to take into account the azimuthal variations of surface-wave velocities. The best method of interpreting sets of fundamental surface-wave data exhibiting a slightly anisotropic behavior with azimuthal variations of velocity is that presented by Montagner and Nataf (1986). As we have already noted in Chapter 2, the first order non-degenerate perturbation theory led these authors, following Smith and Dalhen (1973), to writing down the azimuthal variation of phase velocity as:

$$C(\varphi) = C_0 + \delta\, C(\varphi)$$

where

$$\delta\, C(\varphi) = A_2 \cos 2\,\varphi + A_3 \sin 2\,\varphi + A_4 \cos 4\,\varphi + A_5 \sin \varphi.$$

In these equations, coefficients C_0 and A_n are related to linear combinations of the partial derivatives of a transversely isotropic model with a vertical axis of symmetry (see Chapter 2 and Montagner and Nataf, 1986). Once the local phase velocity C_0 is found together with its 2φ and 4φ azimuthal variations, generally after a former inversion of observed phase velocities has been performed in a regionalization procedure, coefficients C_0, and A_2 to A_5 can be inverted in terms of depth-varying properties. Since the number of parameters to be inverted is rather large, actually 13 linear combinations of elastic coefficients can be measured this way, Montagner and Nataf (1988) proposed to reduce the number of parameters by assuming locally that the structure exhibits an axis of symmetry with unknown orientation. The medium is then supposed to have a hexagonal symmetry with five elastic coefficients plus two angles to orient the axis of symmetry. Such an approach was called "vectorial tomography" by its authors (Montagner and Nataf, 1988) as opposed to scalar tomography in which the seismic velocity is isotropic in each point of the three-dimensional model.

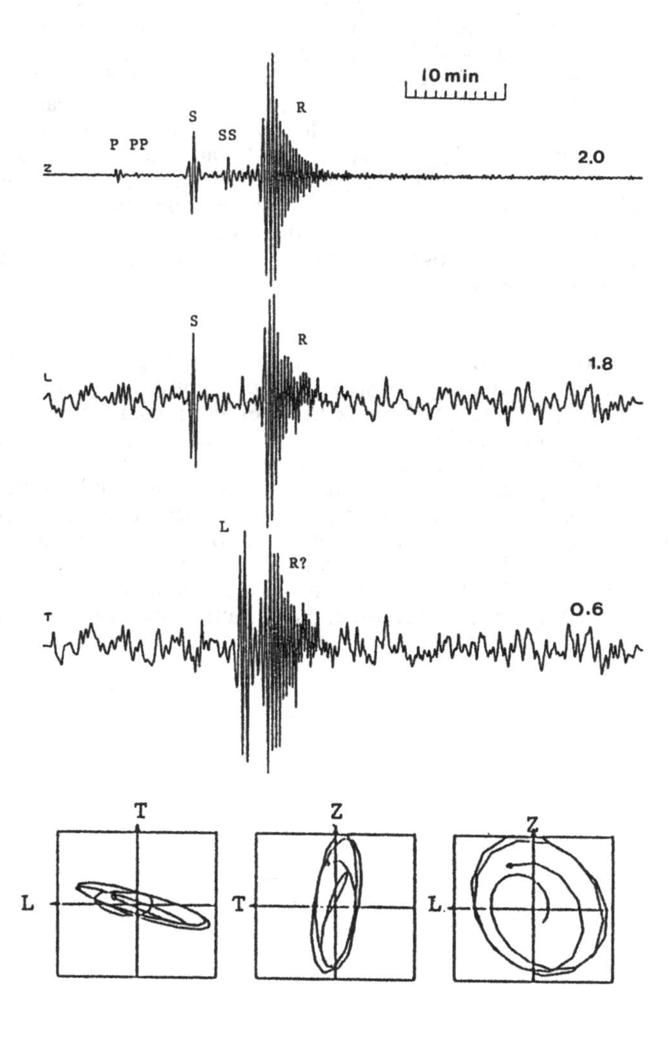

Figure 4-18 Polarization anomaly of Rayleigh waves observed at the Geoscope station Port aux Français, Kerguelen, in the Indian Ocean. L and R denote the Love and Rayleigh waves, respectively. The vertical (Z), radial (L) and transverse (T) components, together with the associated polarization diagrams between the time marks are displayed (after Maupin, 1987). Numbers on the right of each trace give the relative magnification

The first tomographic studies taking into account both lateral variations of velocity and velocity anisotropy were performed in accordance with the conception of vectorial tomography by Montagner and Jobert (1988) who inverted a set of Love- and Rayleigh-wave dispersion curves related to crisscrossing great circle paths in the Indian Ocean. The analysis of Rayleigh-wave velocities along the different paths shows that the dominant azimuthal variation of phase velocity lies in the 2φ term, the variation associated with the 4φ term being poorly resolved. On a global scale, Montagner and Tanimoto (1990, 1991) have provided us with the first global models based on such a type of vectorial tomography algorithms. It is evident from these studies that the best resolution is obtained for the isotropic part of the S velocity which seems to be a robust parameter in surface-wave tomography. Much poorer resolution is obtained on the orientation of the symmetry axis.

Finally, let us briefly comment on the third type of surface-wave observation which can carry some information on the anisotropic properties of the Earth: the polarization anomalies in the surface-wave particle motion. Strong anomalies in the particle motion of Rayleigh waves have been observed in island stations in the Pacific (Kirkwood and Crampin, 1981) and in the Indian Ocean (Maupin, 1987, see Fig. 4-18). In a continental environment let us mention the anomalies of surface-wave polarization reported by Neunhofer and Malischewski (1981) and Lander (1984), (see also Keilis Borok, 1989, p.276). The main problem in interpreting these observations is that it is not easy to separate the effects of lateral heterogeneities beneath the station from anisotropy. Maupin (1987) showed for example that the topography of the Ocean bottom near Kerguelen Island where she measured the polarization anomalies, was irregular and steep enough to strongly affect the particle motion of the Rayleigh waves.

Chapter 5 - Anisotropic structures in the lithosphere

For a short time constant, the lithosphere may be considered as a rigid elastic layer submitted to tectonic stresses. Non-elastic deformations take place in narrow zones located either at boundaries of the tectonic plates or near fault zones within the interior of the plates. Although not proven, it is not impossible that large horizontal shearing deformations take place in the lower part of the crust (e.g. Meissner and Kusznir, 1987). In such a general framework, preferred orientation of minerals may occur in different locations and seismic anisotropy may show up at different scales as a consequence of oriented geodynamic processes during the formation of the lithosphere or, probably more locally, during later tectonic episodes.

As we have seen, anisotropy of the upper crust may depend on a distribution of cracks in the crystalline or sedimentary rocks exposed to regional stresses. Local preferred orientation of minerals in mylonites may also play a role in the vicinity of active tectonic faults (Jones and Nur, 1982; Fountain et al., 1984). Seismic anisotropy of the lower part of the continental crust is not proven, but it may at least exist in connection with the observations of the large scale seismic layering in phanerozoic zones. Subcrustal anisotropy as revealed by azimuthal variations of Pn velocity and splitting of SKS waves is likely to depend on frozen-in orientations of minerals, either related to the initial formation of the lithosphere, as in oceanic areas, or to its past tectonic activity. If these different processes are coherent over a large enough scale, there is no doubt that additional direct evidence for seismic anisotropy in the deep crust and subcrustal lithosphere could rapidly accumulate in the near future, thanks to high quality three-component digital seismic stations.

The anisotropy caused by mineral preferred orientation or the crystalline anisotropy is a typical intrinsic anisotropy which is observable both at very small and long wavelengths. Another type of intrinsic anisotropy is a lithologic anisotropy. According to Crampin et al. (1984a) such a type of anisotropy occurs in sediments when individual grains, which may or may not be elastically anisotropic, are elongated or flattened and these shapes are systematically oriented by gravity or fluid flow when the material is deposited. The transverse anisotropy of clays (Brodov et al., 1984) and shales (Robertson and Carrigan, 1983) is a typical lithologic anisotropy resulting from the orientation of flat mineral grains due to deposition, possibly combined in shales with plastic deformation.

There are several other ordering mechanisms besides sedimentation, which persist over sufficiently large distances to create large-scale anisotropic formations like regional metamorphism and oriented tectonic stresses. Apart from the preferred mineral orientations and distinct foliations, also oriented systems of faults and dykes can be formed. Such large-scale phenomena in combination with a small-scale preferred orientation of minerals and cracks cause the overall anisotropy of seismic wave propagation. All these observations strongly suggest that seismic anisotropy may be more the rule than the exception within the lithosphere.

5.1 Oceanic crust

Most of our knowledge of the thickness and layering of the oceanic crust (Fig. 5-1) has been acquired from the refraction and reflection methods of explosion seismology and from deep drilling. With the increase in resolution and the number of measurements, the early seismic model consisting of a low-velocity layer 1 (sediments in a different stage of consolidation), a 5.0-km/s igneous layer 2 and a 6.8-km/s layer 3

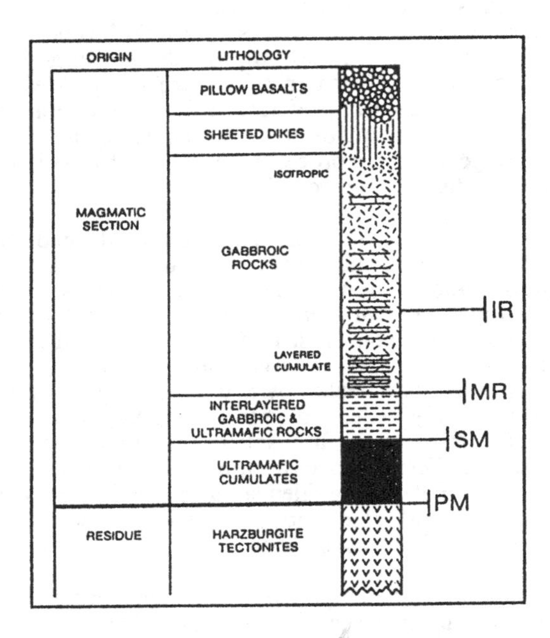

Figure 5-1 Structure of the oceanic crust (Karson and Elthon, 1987). Different Mohorovicic discontinuities can be defined, depending on the type of data: petrological data (PM), seismic reflexions on the discontinuity (MR), seismic refraction data (SM), or intracrustal seismic reflections (IR). (Geological Society of America, copyright ©1987).

evolved into more detailed models (e.g. Fig. 5-1). Both seismic and drilling results indicate a substantial amount of inhomogeneity and anisotropy in the upper crust.

Anisotropy of marine sediments results primarily from bedding the interleaving of thin parallel layers of contrasting mineralogical composition (Carlson et al., 1984), which is described by transverse isotropy with a vertical axis of symmetry. Berge et al. (1991) described such a type of anisotropy in the top 50 m of marine sediments composed of interbedded silty clays, clays and sand near the coast of New Jersey. They estimated all five elastic stiffnesses for a transversely isotropic medium and found that the top 10 m of the sediments exhibit 12-15 % of anisotropy.

The structure of the oceanic crust, in general, varies with direction (Stephen, 1981). Besides the symmetry inherited from the spreading process, this phenomenon is explained by linear seafloor topography and well oriented faulting and fracturing of the oceanic basement (Ballard and van Andel, 1977). Hence, models of the formation of the oceanic crust include aligned lava flows and oriented dyke systems in the shallowmost crust (Cann, 1974; Sleep and Rosendahl, 1979). This leads to observations of seismic anisotropy in the oceanic basement.

Stephen (1981) reported observations of seismic anisotropy in the upper oceanic crust in the west Atlantic, demonstrated by the polarization analysis of three-component borehole seismometer data. The anisotropic effects were too small to be resolved by means of the conventional travel-time analysis, however, he observed decoupling of shear waves into quasi-horizontally polarized and quasi-vertically polarized phases which propagated at different velocities (horizontal polarizations arriving after vertical polarizations). Unfortunately, the data were only obtained at azimuths parallel and perpendicular to the predicted spreading direction. White and Whitmarsh (1984), Shearer and Orcutt (1985) reported observations of anisotropy in the upper oceanic crust based on azimuthal variations of compressional wave velocities.

In a more detailed study, Stephen (1985) showed from compressional wave arrival times at short ranges and shear-wave particle motions that seismic anisotropy was present in the upper crust in the equatorial east Pacific (Fig. 5.2). Both data sets are consistent with a transversely isotropic upper crust with a horizontal symmetry axis. The slow direction of compressional waves is within 20° of the predicted spreading direction for the crust at the site. Compressional wave velocities are 4 km/s in the slow direction and 5 km/s in the fast direction. The observed anisotropy, which seems to be confined to the upper 500 m of the crust, is most probably caused by the preferred orientation of large-scale fractures and fissures created in early stages of crustal development by near-axis extensional processes and normal-block faulting. Apparently, the effects of the remanent fissures and faults diminish with depth.

Francis (1976) determined anisotropy of P and S waves from the records of ocean bottom seismographs operating in the medium valley of the Mid-Atlantic Ridge. The recordings of locally occurring earthquakes showed an unusually high ratio of compressional to shear velocity which is explained by a high fissure porosity near the axis of the ridge. Thus the propagation of waves in the upper part of the crust with a strong preferred orientation of faults and fissures is anisotropic. In addition, shear-wave birefringence will occur for all directions of propagation except for the direction perpendicular to the fissures, i.e. in the spreading direction. The author concludes that the anisotropic properties will tend to diminish as the crust moves away from the ridge axis.

Anisotropy in the upper oceanic crust is widely assumed to result from parallel vertical fractures and could thus exhibit a transverse isotropy with a horizontal symmetry axis. However, there are also horizontal fractures, which together with interbedded flows and breccia units, may introduce an associated transverse isotropy with a vertical symmetry axis. These two fracture sets yield orthorhombic symmetry

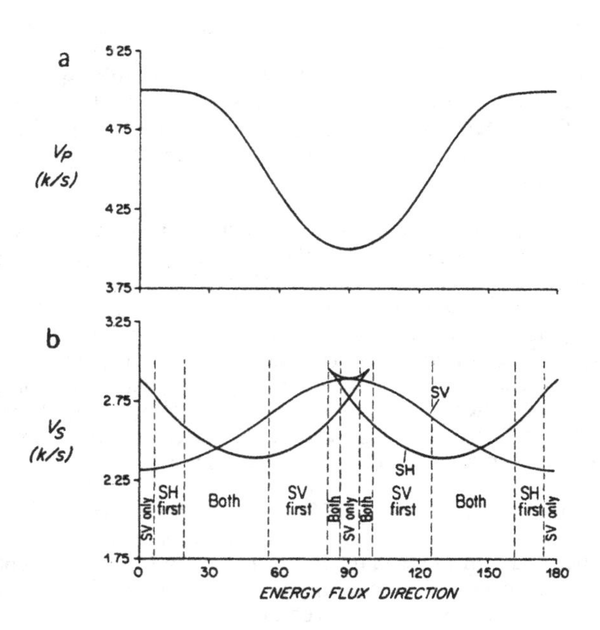

Figure 5-2 Anisotropic model explaining P-wave arrival times and S-wave particle motions observed at Deep Sea Drilling Project in the equatorial east Pacific (Stephen, 1985). Curves show the magnitudes of group velocities for compressional (a), horizontally polarized shear (S_H), and vertically polarized shear (S_V) waves (b) as a function of azimuth. The spreading direction at the site is North-South. (AGU, copyright © 1985).

(Hood et al., 1988).

As Bibee and Shor (1976) suggested, the azimuthal variations of seismic velocities may be affected by a number of lateral heterogeneities related to the crustal age, spreading rate and tectonic setting. The authors have found that seismic velocities in the oceanic crust do not vary significantly with the azimuth, age or geographic region and that all variations are probably local phenomena. In contrast to crustal velocity variations, mantle velocities beneath the oceans exhibit a widespread and more systematic anisotropy.

5.2 Subcrustal lithosphere under oceans

The oceanic lithosphere is created beneath the midoceanic ridges where a transfer of mass takes place due to upwelling of an asthenospheric material and its cooling and solidification. As the adjacent plates diverge, hot mantle rock ascends to fill the gap. The cooling rock accretes to the base of the spreading plates and as the plates move away from the ocean ridge they continue to cool and thicken. At the interface between the lithospheric plate and the asthenosphere the anisotropic mineral alignment is formed, namely the alignment of olivine crystals, which leads to the observed seismic anisotropy.

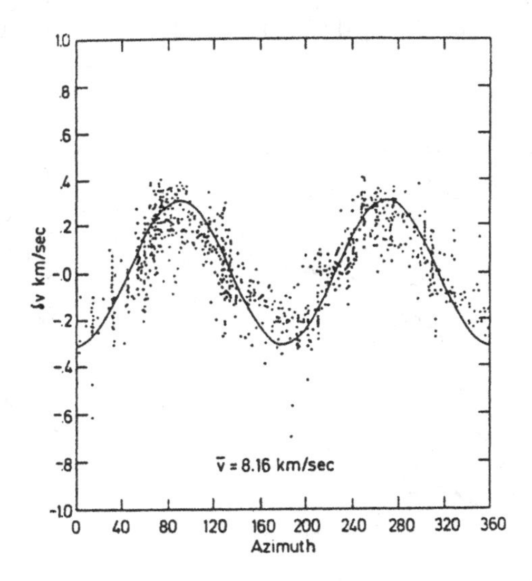

Figure 5-3 Azimuthal dependence of Pn-wave velocities in the uppermost mantle near Hawaii (after Morris et al., 1969). (AGU, copyright ©1969).

5.2.1 Observations

Hess (1964) first published evidence that the uppermost mantle in the northwest Pacific was anisotropic to compressional wave propagation and suggested that the anisotropy of about 7% was due to a preferred orientation of olivine crystals. His conclusions were based on observations of Pn velocity variations with azimuth made by Raitt (1963) and Shor and Pollard (1964). The fastest Pn velocities were approximately east-west and parallel to trends of fracture zones which were presumably fossil traces of the paleospreading of the Pacific plate (Fig. 5-3). The slowest direction was perpendicular to that direction and parallel to magnetic lineations. Similar observations were made in various regions of the Pacific Ocean (Morris et al., 1969, see Fig. 5-3; Raitt et al., 1969; Keen and Barrett, 1971; Shor et al., 1973; Snydsman et al., 1975; Malecek and Clowes, 1978; Shimamura et al., 1983; Shearer and Orcutt, 1985) and the Indian Ocean (Shor et al., 1973). Refraction measurements in the vicinity of the

Mid-Atlantic Ridge (Keen and Tramontini, 1970) suggest the presence of seismic anisotropy in the upper mantle, however, the scatter in the velocities resulting from the rough topography makes it difficult to demonstrate clearly its existence. The azimuthal anisotropy of Pn velocities in the oceanic lithosphere can amount to 10%, but it generally ranges from 3% to 8% with the fastest directions oriented along the fossil directions of the paleo-plate spreading (Fig. 5-4).

The anisotropy, originally established from azimuthal differences of Pn waves was confirmed by observations of other types of seismic waves. As was shown, e.g. by Crampin (1977), the most straightforward feature of wave propagation through an anisotropic medium is the splitting of shear waves which propagate with different polarizations and at different velocities. On the two rotated seismograms the two waves are separated, and from the velocity difference the anisotropy is evaluated.

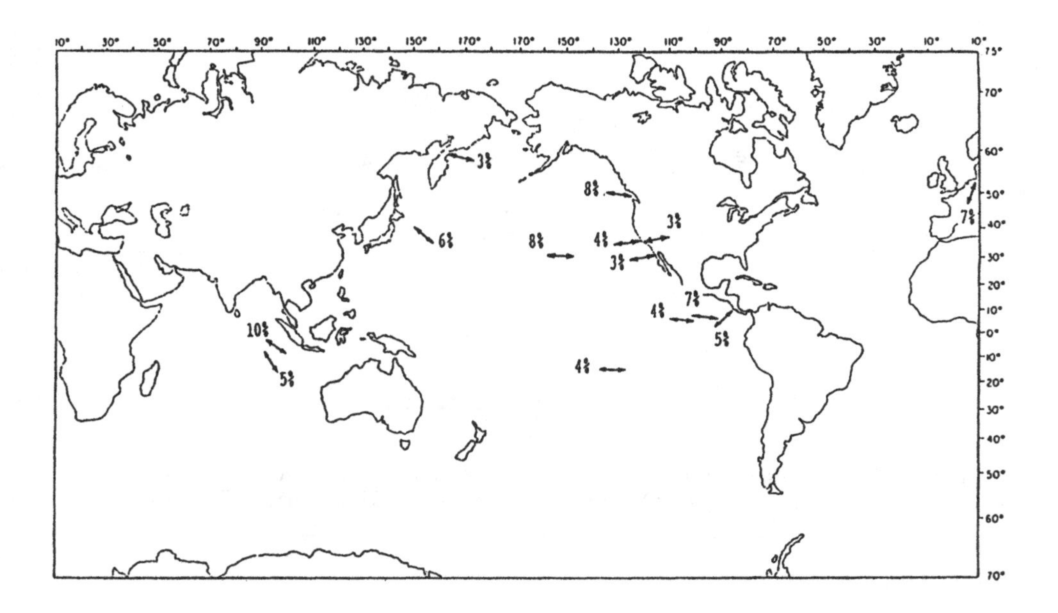

Figure 5-4 Observed anisotropy of Pn velocities in the uppermost mantle. Arrows are parallel to fast directions (Christensen, 1984). (With permission of the Royal Astronomical Society, copyright © 1984).

Using data from the three-component seismographs of the World-Wide Standardized Seismograph Network, Ando (1984) established ScS polarization anisotropy from short-period seismograms of intermediate-depth and deep earthquakes recorded at 24 stations around the Pacific Ocean. The core-reflected ScS waves propagate vertically in the mantle with two mutually orthogonal directions of linear particle motions. The polarizations of the faster shear waves are generally parallel to plate motion near the receiver points (Fig. 5-5). The average arrival-time difference is 1.0 +/-0.4 s, corresponding to a velocity difference of 4% in a 100 km thick anisotropic layer of the upper mantle.

Ansel and Nataf (1989) analysed polarizations of SKS, ScS and S waves on seismograms of the Geoscope broadband network. They found large anisotropies at all the three stations in the Pacific Ocean, however, contrary to the fast Pn velocities oriented along the spreading direction, their results indicate that the polarization of the faster S wave is almost orthogonal to the present or fossil spreading direction. According to the authors this observation could indicate that the mantle beneath oceanic islands is different from the "normal" oceanic mantle possibly due to a hot-spot related flow which strongly reorganized the base of the lithosphere and the mantle beneath it. Another explanation can be seen in a "layered" anisotropic structure of the oceanic lithosphere which will be discussed later in this chapter.

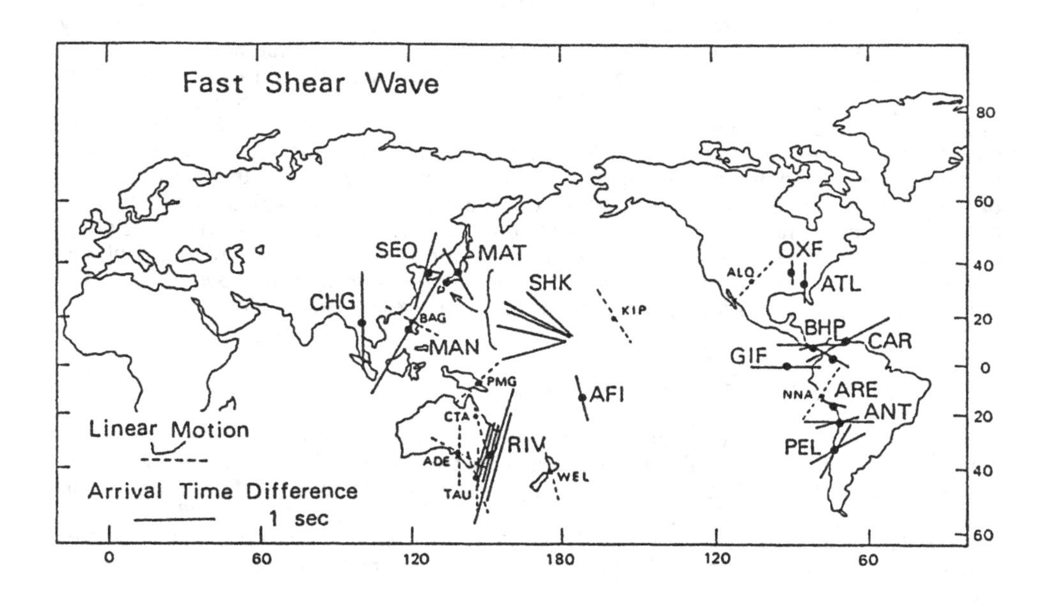

Figure 5-5 Polarizations of faster split ScS waves (Ando, 1984). The solid lines show the direction of polarization of the faster split shear wave in the horizontal plane, and the dashed lines with a smaller station code show the linear polarization of the particle motion where there is no splitting. The length of the solid lines is proportional to the time difference between the two split ScS waves. (With permission of the Center for Academic Publications, Japan, copyright © 1984).

Information on shear anisotropy is also provided by Love- and Rayleigh-wave data since surface waves are sensitive primarily to shear-wave velocity. The anisotropy in both the lithosphere and the asthenosphere has been inspected by surface-wave studies for a long time. The main evidence comes from the fact that the Love- and the Rayleigh -dispersion curves which are constructed for the same region are not generally compatible with an isotropic S-velocity model of the upper mantle. This Love-Rayleigh -wave discrepancy can be accounted for either by a transversely isotropic model if no azimuthal variations of the velocity are observed, or by a fully anisotropic model (see section 2.5). The problem with classical surface wave observations is that a rather loose depth resolution is obtained when only the fundamental modes of the Love and Rayleigh waves are used. To illustrate this problem, we can mention Schlue and Knopoff (1976) who proposed from such data an isotropic lithosphere overlying a transversely isotropic asthenosphere, while Mitchel and Yu (1980) propose a reverse model.

A more recent analysis of the Pacific Ocean data based upon fully anisotropic models of the upper mantle suggests that a small azimuthal variation of the surface wave velocity is present and the anisotropy may extend over the whole lithosphere and most likely to at least 200-300 km (Montagner, 1985; Kawasaki, 1986). Furthermore, in the Pacific Ocean the fast direction of the Rayleigh waves coincides with the present-day motion direction of the Pacific plate so that the most plausible explanation is that the whole lithosphere may be anisotropic and, in accord with the Pn observations, its anisotropy is due to the alignment of olivine crystals which are oriented with the a-axis in a direction parallel to the plate motion (Nishimura and Forsyth, 1989). In regions where directions of the plate motion changed drastically in the past, like in the northwest Pacific, the observed direction of the fast Rayleigh waves coincides with the past motion of the plate. A frozen-in orientation of the olivine crystals might explain such a type of observation, as in the case of the Pn anisotropy. Azimuthal anisotropies of Pn velocity and of shorter period Rayleigh waves require anisotropy within the oceanic lithosphere. The azimuthal anisotropy of about 2% for Rayleigh waves extending to periods of 250 s would also require anisotropy in the asthenosphere as we will see in Chapter 6.

5.2.2 Causes and origin of anisotropy

Geophysical and petrological constraints restrict the dominant mineralogical composition of the subcrustal oceanic lithosphere to olivine and pyroxene. Dick (1987) made modal analyses of 266 peridotite samples dredged at 6 ocean ridges and gave an average composition of 74.8 ±5.3% olivine, 20.6 ±3.7% enstatite, 3.6 ±2% diopside, 0.5 ±0.2% Cr-spinel and 0.5 ±1.2% plagioclase. Thus we may assume that velocities and a preferred orientation of olivine and pyroxene are most probably responsible for the observed seismic anisotropy.

A detailed review of the formation of the anisotropy in the upper mantle peridotites was published by Nicolas and Christensen (1987). On the basis of laboratory measurements of single crystal elastic constants of olivine (Verma, 1960) and compressional wave velocities and their anisotropy determined for peridotite by Birch (1960), Hess (1964) recognized already a long time ago the real cause of the velocity anisotropy of the oceanic lithosphere in the preferred orientation of olivine crystals. He ascribed the alignments of olivine (010) slip planes to plastic flow along transform faults. Francis (1969) proposed that the anisotropy took its origin in the asthenospheric flow at oceanic ridges, which oriented [100] olivine slip lines at high angles to the ridge trend. Experimental studies on the development of preferred orientations of olivine aggregates (Ave Lallemant and Carter, 1970; Nicolas et al., 1973) demonstrated that distinct lattice preferred orientations were obtained for moderate strains, which were related to plastic flow (Nicolas et al., 1973). The volumetric predominance of olivine in peridotites explains its dominant influence on the anisotropy. The largest compressional velocity coincides with the [100] axis which is the dominant slip line at a high temperature and the smallest velocity with the [010] axis which is normal to a common slip plane (Fig. 5-6). Consequently, the compressional-wave anisotropy in olivine aggregates is directly related to the plastic flow directions (Nicolas and Christensen, 1987). This is not true for the shear velocities in olivine (see Chapter 3), whose pattern is not simply related to the slip orientation.

Pyroxenes have a diluting or negative effect on the anisotropy (Fig. 5-7, Nicolas and Christensen, 1987), other minerals of peridotites (garnet, spinel, serpentine) have simply a diluting effect. Although the anisotropy of the subcrustal oceanic lithosphere may originate from effects other than the preferred orientation of anisotropic minerals, like the stratification of the upper mantle, it has been generally accepted that the alignment of olivine crystals has a dominant effect.

Studies of peridotite samples from volcanic xenoliths and orogenic massifs show that their structures reflect high to very high temperature flow conditions favoring the development of the strongest preferred orientations of olivine (Nicolas and Poirier, 1976). Medium to low temperature deformations are confined to local areas, such as margins of mantle diapirs or shear domains ascribed to transform faults (Nicolas and Christensen, 1987). Christensen (1984) concluded that the symmetry of ultramafites varied from orthorhombic to axial with the olivine [100] axes maxima being subparallel to the spreading directions in the oceanic upper mantle. At the high and very high temperature conditions (>1100°C) the [010] axes maxima, which are the low-velocity directions, are approximately vertical (Fig. 5-6) and the resulting symmetry is orthorhombic. Such symmetry is typical for ultramafites with a high olivine content (e.g. Babuska and Sileni, 1981). The orthorhombic model of the subcrustal oceanic lithosphere was suggested by Nicolas and Poirier (1976).

Under medium (1000°C) to high temperature conditions the directions of the lowest velocity ([010] olivine axes) and of the intermediate velocity ([001] axes) form partial girdles in a vertical plane which is perpendicular to the [100] axes oriented in a sea-floor spreading direction (Fig. 5-6, Nicolas and Christensen, 1987) and the resulting symmetry of the structure is close to transversely isotropic with the horizontal symmetry axis. Such a model is favored, e.g. by Kawasaki and Konno (1984), who examined surface wave group velocities in the Pacific plate and found an azimuthal anisotropy in the Rayleigh wave propagation but little anisotropy in Love waves. On the basis of petrofabrics of dunite from Hidaka (Japan) and peridotites from the ophiolite complex of the Bay of Islands (Newfoundland), they derived a model with [100] olivine axes oriented along flow directions (normal to the oceanic ridge) and the remaining crystallographic axes of olivine randomly oriented in the plane perpendicular to [100] axes. Strictly speaking, most ultramafic rocks studied by various authors, show fabrics which are closer to the orthorhombic symmetry (or even monoclinic) than to the transversely isotropic symmetry.

T domains	dominant slip systems	Fabrics strength, type	Relation with flow plane (horizontal) and flow line (arrow).
⩾1250°C (hypersolidus)	(010) [100] (001) [100]	- extreme - point maxima partial girdles	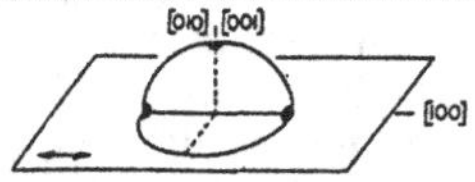 [010], [001]; [100]
>1100°C (high-T)	(010) [100]	- strong - point maxima	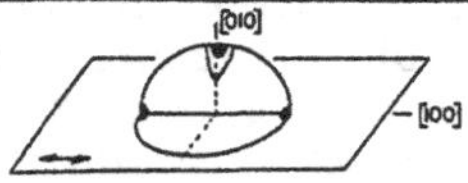 [010]; [100]
~1000°C (medium-T)	(0kl) [100]	- distinct - point maximum and girdle	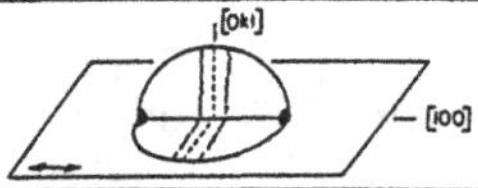 [0kl]; [100]
700-1000°C (low-T)	(0kl) [100] (010) [001]	- absent or weak - double diffuse girdle	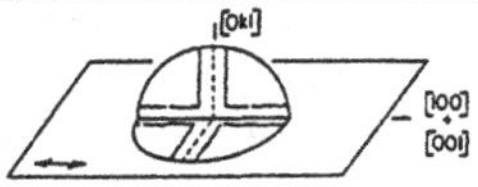 [0kl]; [100] + [001]

Figure 5-6 Olivine slip systems and flow orientations in the function of temperature for large shear strains (Nicolas and Christensen, 1987). (AGU, copyright, © 1987).

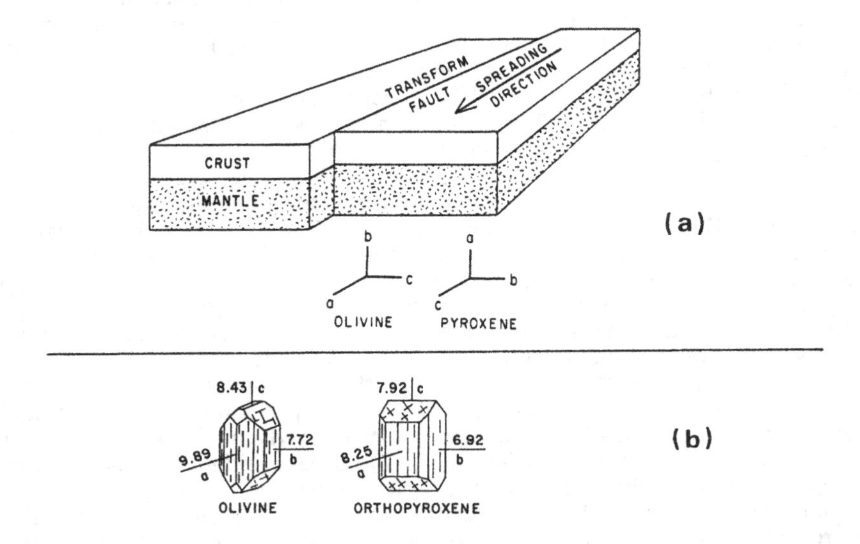

Figure 5-7 The ideal orientations of olivine and orthopyroxene in the oceanic upper mantle. Below are P-wave velocities in km/s along the crystallographic axes of both minerals (after Christensen, 1984). (With permission of the Royal astronomical Society, copyright © 1984).

The study of oceanic lithosphere has been aided by the investigation into characteristic rock sequences known as ophiolites. They usually occur in collisional mountain belts and their association of deep sea sediments, basalts, gabbros and ultramafic rocks suggests that they originated in the oceanic lithosphere and were subsequently thrust up into their continental setting by a process known as obduction (Ben-Avraham et al., 1982; Kearey and Vine, 1990).

Laboratory investigations of ultramafic rocks which represent upper mantle materials and which were originally parts of an ancient oceanic lithosphere (see Chapter 3) provide a basis for interpretations of the observed seismic anisotropies. It is especially desirable to compare the mantle anisotropy with laboratory measurements on multiple samples collected over large ultramafic massifs (Christensen, 1978; Peselnick and Nicolas, 1978; Christensen and Salisbury, 1979). Within these massifs the degree of preferred mineral orientation and symmetries often vary significantly from one locality to another; however, the directions of the maxima of corresponding olivine axes are surprisingly consistent over large areas (Christensen, 1984). This is true for ultramafic massifs of Oman, Turkey, Cyprus, the Alps, Newfoundland, the United States and New Zealand, which display remarkable similarities, despite their different ages and a wide range in the geographic distribution. These observations suggest that the seismic anisotropy has been a property of the upper mantle throughout much of the geological time and is world-wide in occurrence (Christensen, 1984).

Christensen (1984) compiled the Pn anisotropies determined by seismic refraction studies in the uppermost mantle of the Pacific and Indian Oceans (Fig. 5-4) and concluded that the observed anisotropies between 3% and 10% agreed with the compressional wave anisotropy calculated for ultramafic massifs from several ophiolite complexes. Average P-wave anisotropies of olivine-rich ultramafites which also include xenoliths from volcanic rocks are about 8-9% (see Chapter 3); and exceptionally they attain 15% in dunite (Babuska, 1972b). The higher anisotropy determined in rock samples is most probably due to the fact that laboratory measurements determine the anisotropy three-dimensionally, whereas the seismic refraction data provide mainly a two-dimensional picture of anisotropy from subhorizontally propagating Pn waves. Moreover, systems of profiles can miss some of the velocity extremes in the horizontal plane. In the case of the orthorhombic symmetry with the [010] olivine axes oriented vertically (Fig. 5-7), the Pn velocities do not allow us to determine a maximum anisotropy, because the velocity minimum is vertical. Therefore, it is very probable that real P-velocity anisotropies of the oceanic subcrustal lithosphere are larger than those determined by the seismic refraction experiments.

5.2.3 Depth range of anisotropy in the oceanic lithosphere

The oceanic lithosphere is thought to be a rigid slab with a thickness of about 70 to 100 km (Asada, 1984). Ando (1984) estimated the anisotropic layer derived from ScS polarization anisotropy and the differences in arrival times between two split waves reflected from the mantle-core boundary to be about 100 km thick. It is an important question whether the velocity anisotropy exists over the entire thickness of the oceanic lithosphere. Another question is the possibility that the lithosphere might be thinner if seismic anisotropy is taken into account (Regan and Anderson, 1984). This point is discussed in Chapter 6. We will see that there is in fact no clear necessity to reconsider the classical estimate of the oceanic lithospheric thickness.

Shimamura et al. (1983) conducted long-range explosion experiments in the northwestern Pacific where the oldest oceanic lithosphere exists (Fig. 5-8). The length of profiles up to 1800 km and a wide azimuthal coverage enabled the authors to find out that the azimuthal anisotropy varies between 4 and 7% and extends at least from the depth of 40 km to 140 km beneath the sea bottom. The azimuth of the maximum compressional velocity is N 150-160° E. This direction coincides with that of the past spreading of the Pacific plate 40 million years ago, whereas it differs from the present direction of the plate motion by about 30°. They conclude that the degree of anisotropy is likely to vary with depth and the subcrustal lithosphere of the Pacific plate is basically two-layered with the interface at 50 - 60 km beneath the sea floor.

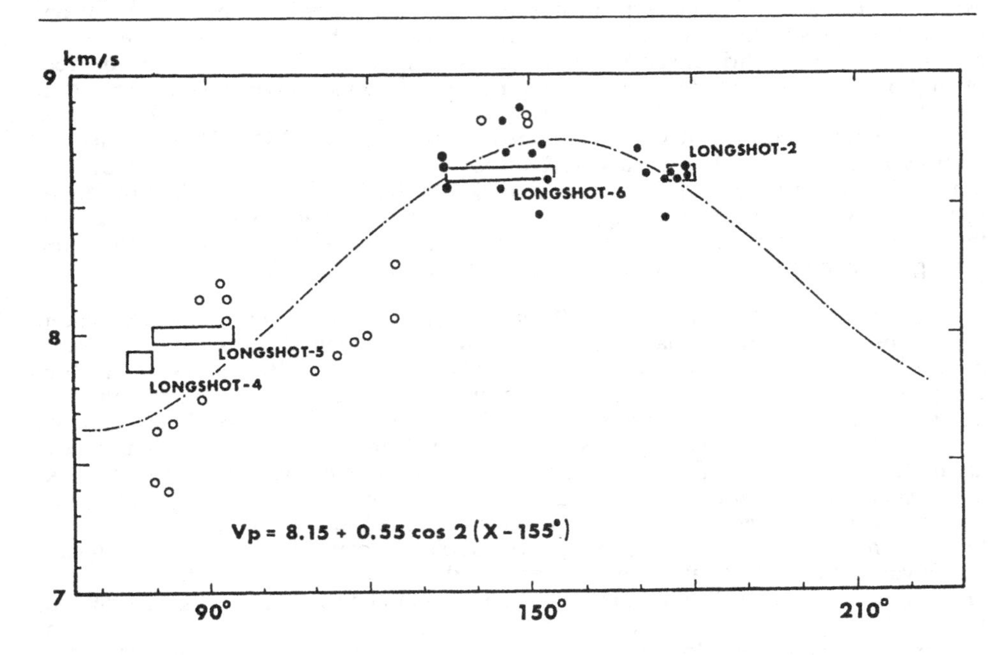

Figure 5-8 P-velocity anisotropy in deeper parts of the oceanic lithosphere beneath the northwestern Pacific (Shimamura et al., 1983). Data are from Longshot experiments (rectangles) and natural earthquakes (open and solid circles). The azimuth of the fast velocitiy (150°) corresponds to the fossil seafloor spreading (Nishimura and Forsyth, 1989). (With permission of Elsevier Science Publisher, copyright © 1983).

Asada (1984) reported a very clear refraction arrivals recorded in the northwestern Pacific in a distance range from 800 km to 1000 km. Besides confirming that the velocity maximum is perpendicular to the magnetic lineations, the author determined the mean Pn velocity at 8.15 km/s and a difference between velocity extremes at ±0.55 km/s, and concluded that the P-velocity anisotropy probably exists throughout the entire depth of the lithosphere.

As mentioned at the beginning of Chapter 5, the maxima of Pn velocities in the oceanic lithosphere are normal to mid-ocean ridges, i.e. normal to magnetic lineations, thus reflecting the orientation of the mantle flow within the young oceanic lithosphere. Since the maximum velocity in an old oceanic lithosphere is still normal to magnetic lineations but considerably differs from the present plate motion (Shimamura, 1984), the anisotropy must have been created within the young lithosphere near the mid-ocean ridge and the older lithosphere most probably retains that original frozen-in pattern of anisotropy.

To interpret a depth-dependent velocity anisotropy observed by Shimamura et al. (1983), Ishikawa (1984) proposed a layered anisotropic plate thickening model. The anisotropic alignments in such a model are assumed to form at the interface between the lithosphere and the asthenosphere. The olivine orientations vary with depth and the direction of [100] olivine axes in each layer represents the plate absolute motions at the time this layer was accreted to the lithosphere (Fig. 5-9). The difference in formation time of the lithosphere near and far away from the mid-ocean ridge can explain the different anisotropy in the upper and lower layers of the Pacific plate. Besides a turn of olivine crystals around [010] axes, we have to assume anisotropy variations across the oceanic plate, due to changes in the amount of oriented crystals in dependence of the changing physical conditions at the bottom of the cooling oceanic plate and a speed of flow in the asthenosphere.

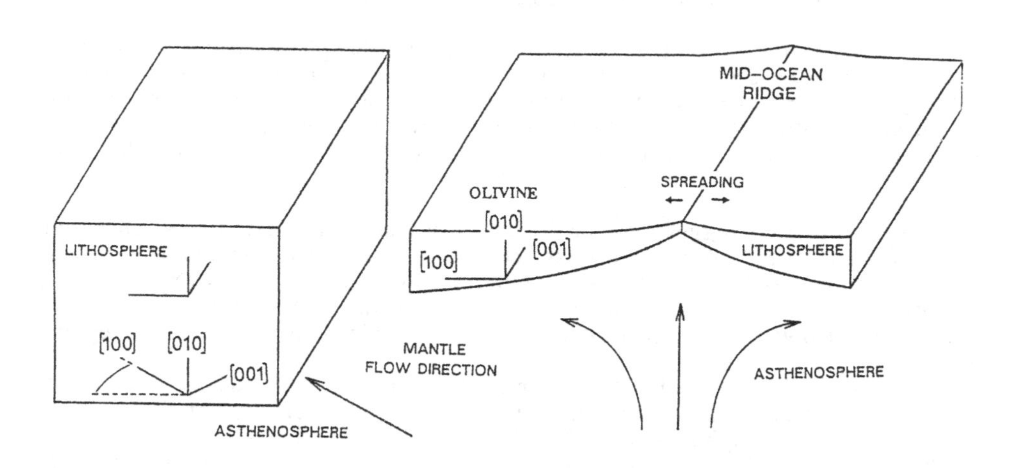

Figure 5-9 Schematic representation of the variable orientation of seismic anisotropy in different depths of the oceanic subcrustal lithosphere based on observations by Shimamura et al. (1983) and the anisotropic plate thickening model by Ishikawa (1984). Olivine crystals are aligned at the lithosphere-asthenosphere transition with [100] axes in the direction of the mantle flow; with a change of the flow orientation away from the mid-ocean ridge, also the velocity maximum along [100] axes has changed its orientation in a newly formed deeper part of the lithosphere.

Seismic anisotropy of the subcrustal oceanic lithosphere has been proven by many independent observations. The anisotropy is caused by the frozen-in preferred orientation of olivine crystals which resulted from the plastic flow beneath the base of the newly formed oceanic lithosphere. Maximum P-wave velocities are normal to mid-oceanic ridges thus reflecting orientations of olivine a-axes along the mantle flow. The anisotropy is probably a typical property of the entire thickness of the subcrustal lithosphere, though the orientation of velocity extremes changes both laterally and vertically in dependence on the orientation of the flow in the asthenosphere in different stages of the lithosphere formation.

5.3 Subduction zones

Subduction zones belong to the most dynamic tectonic environments on the Earth, where large accretionary prisms are formed of sediments scraped from the subducting oceanic lithosphere. Physical properties in accretionary prisms change during their transition from high-porosity basinal sediments to deeply buried low-porosity metamorphosed rocks. Seismic sections of accretionary complexes not only image the deformation structures, but velocity distributions can be interpreted in terms of the *in situ* physical properties (Langseth and Moore, 1990). Deformation mechanisms of different scales (slaty cleavage, schistosity, recrystallization) produce systems of oriented structures which, together with systems of oriented faults, produce a highly anisotropic medium for seismic wave propagation (Fig. 5-10).

In subduction zones the oceanic subcrustal lithosphere, whose anisotropic structure was discussed in the previous section, is descending into the upper mantle. A sinking lithosphere slab represents an inhomogeneity in the surrounding mantle and influences the propagation of seismic waves. An important question arises as to what extent this propagation is also affected by a possible anisotropy of the subducting lithosphere.

One of the first suggestions for the anisotropy under the island arcs came from Sugimura and Uyeda (1967). They assumed that the preferred orientation of olivine crystals accounted for the observed radiation patterns of deep earthquakes, and suggested that the preferred orientation might be the result of the regional compressive stress associated with the upper mantle current under the Izu-Mariana Island arc region.

Cleary (1967) observed a large azimuthal variation of travel times from nuclear explosion Longshot and explained this feature by a strong velocity anisotropy of the high-velocity slab dipping beneath the Aleutian arc. The author sees the most likely explanation in the anisotropic properties of oriented olivine grains. He arrived at an important conclusion that if such a structure was common to arc regions, it represented a source of error in the determination of earthquake locations.

Fukao (1984) observed ScS polarization anisotropy (with an average arrival time delay of 0.8 ±0.4 s) and consistent shear-wave polarizations at 20 stations of the Japan Meteorological Agency. Ando et al. (1983) analysed waveforms of deep earthquakes in the subducting Pacific Plate beneath Honshu and detected clear split arrivals of shear waves. They found that polarizations in Japan vary with space even within a distance smaller than 100 km and concluded that anisotropic bodies seem to be distributed

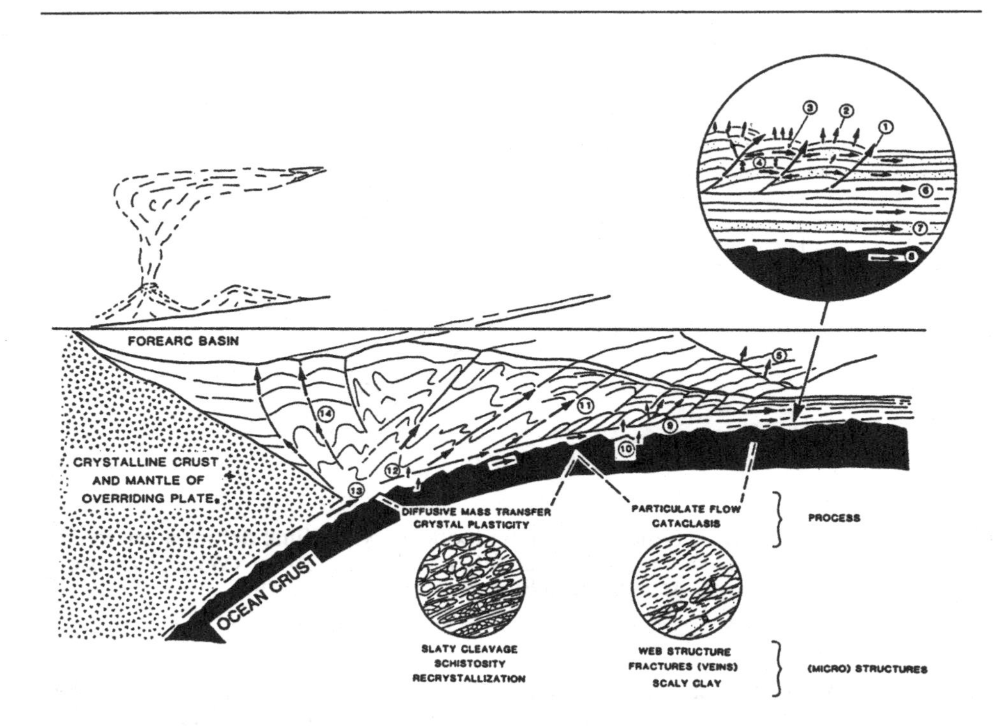

Figure 5-10 Deformational mechanisms and possible fluid flow paths (circled numbers) in an accretionary prism (Langseth and Moore, 1990). Deformational mechanisms, illustrated in the two circles at the bottom of the figure, which can range from grain boundary sliding at shallower levels to diffusive or crystal-plastic processes at depth, produce a medium which is probably characterized by seismic anisotropy. (AGU, copyright © 1990).

irregularly throughout the wedge portion of the upper mantle above the descending Pacific plate beneath central Japan. This interpretation is inconsistent with the uniform distribution of polarizations over Japan observed by Fukao (1984). This problem deserves further investigations with a more complete set of seismological data (Ando, 1984).

Hirahara and Ishikawa (1984) extended the method by Aki et al. (1977) and developed an inversion for estimating the three-dimensional anisotropic P-wave velocity structure beneath the seismological stations in southwest Japan. Although the anisotropic velocity perturbations are not so well resolved as the isotropic ones, the authors established an anisotropic propagation of seismic waves within the descending

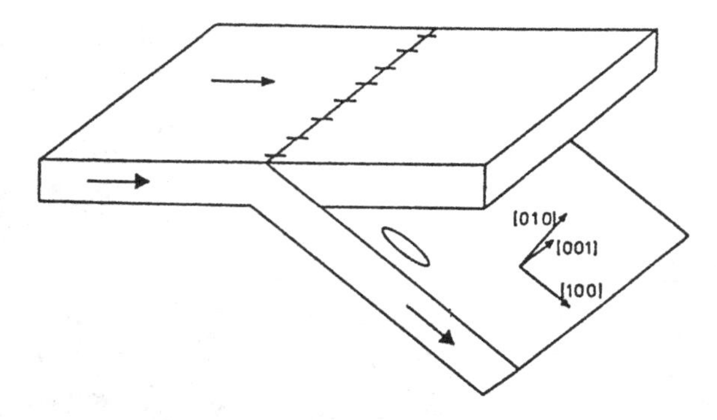

Figure 5-11 A possible olivine orientation produced by the final deformation in the mantle above the sinking slab (after McKenzie, 1979). (With permission of the Royal Astronomical Society, copyright © 1979).

Pacific and Philippine Sea plates. The fast directions of P velocities seem to be perpendicular to the magnetic lineations left in the floors near the trenches. The explosion experiments in the northwestern Pacific also showed the high-velocity directions of P waves perpendicular to magnetic lineations in that region. These results suggest that the anisotropic structures which were formed before subducting are preserved in the descending lithosphere beneath Japan.

As a preferred crystal orientation once formed is not easy to destroy (McKenzie, 1979), it is probable that the olivine orientation and a corresponding seismic anisotropy inherent to the oceanic subcrustal lithosphere are preserved in the subduction zones down to a depth of at least 400 km, where P-T conditions are favorable for a phase transition of olivine from the orthorhombic to the spinel structure. It is probable that due to a well-known delay in changes of P-T conditions inside subducting slabs the original olivine orientation and the anisotropy can be preserved to greater depths than 400 km. The anisotropic propagation within the subducting oceanic lithosphere should have important consequences on 3-dimensional inversions of P residuals (e.g. Spakman et al., 1988). The rays propagating within the subducting slabs are greatly accelerated not only because of the thermal effects, but also because of the anisotropy. In this way cumulative effects of high-velocity directions along the subducting slabs can produce false high-velocity inhomogeneities in tomographic inversions of P residuals showing a much deeper penetration of the subducting slabs.

When we think of a model of olivine preferred orientation proposed by Nicolas and Poirier (1976) and Christensen (1984), discussed in the previous section, then olivine a- and c-axes would lie in the plane of the subducting plate and b-axis (direction of the minimum P velocity) would be approximately perpendicular to the subducting plate. This means that due to the anisotropy the P velocities are always larger along the subductions than across them.

Apart from the anisotropy created in early stages of the development of the oceanic lithosphere, there might be another mechanism producing the anisotropy in the subduction zones. McKenzie (1979) showed that the strain induced by subduction above the sinking slabs was much larger than that beneath ridges, and should produce important anisotropy with the highest P velocity (olivine a-axes) oriented in the dip direction of the sinking slab (Fig. 5-11). Such a hypothetical preferred olivine orientation would enhance a velocity difference between directions along and across a subduction zone.

5.4 Continental crust

Compared with oceanic regions, the crust of continents shows a more complicated picture. The main reason is that the oceans are relatively young, 95% of the Earth's history is imprinted in the continental lithosphere. The age of the oldest parts of the continental crust is at least 3.8×10^9 years (Anderson, 1989) and the mean age of the continental crust is 1.5×10^9 years (Jacobsen and Wasserburg, 1979).

Unlike the oceanic crust, which appears to form continuously by partial melting from a source of a more or less uniform composition, the origin of the continental crust is obscure and its evolution seems to have been complex. Geological and geochemical examinations of continental rock sequences clearly suggest that many of these originated in oceanic settings through plate-tectonic processes. Oceanic rocks were incorporated into continents by such processes as island arc or continental collisions and by accretion of oceanic and volcanic rocks at noncollisional convergent boundaries.

In Chapter 3 we demonstrate an extreme variability of P- and S-wave velocities in samples of crustal rocks, and a high intrinsic elastic anisotropy of some sedimentary and metamorphic rocks. In the upper crust the overall pattern of anisotropy is moreover affected by oriented systems of cracks of various dimensions. In this part we want to discuss how the seismic wave propagation can be affected by the anisotropy in different regions of the continental crust.

5.4.1 Upper crust

The layering of sediments has long been recognized as a source of seismic anisotropy (e.g. Riznichenko, 1949; Postma, 1955). Seismic velocities in horizontally layered media are independent of the azimuth of propagation. Such a transverse isotropy with a vertical axis of symmetry (see Chapter 2) can be combined with a transverse isotropy characterized by a horizontal axis of symmetry due to parallel vertical cracks, which leads to an orthorhombic symmetry (Leary et al., 1990).

Direct observations of azimuthal seismic anisotropy in sedimentary layers are now quite frequent. We will report here on one clear evidence coming from a vertical seismic

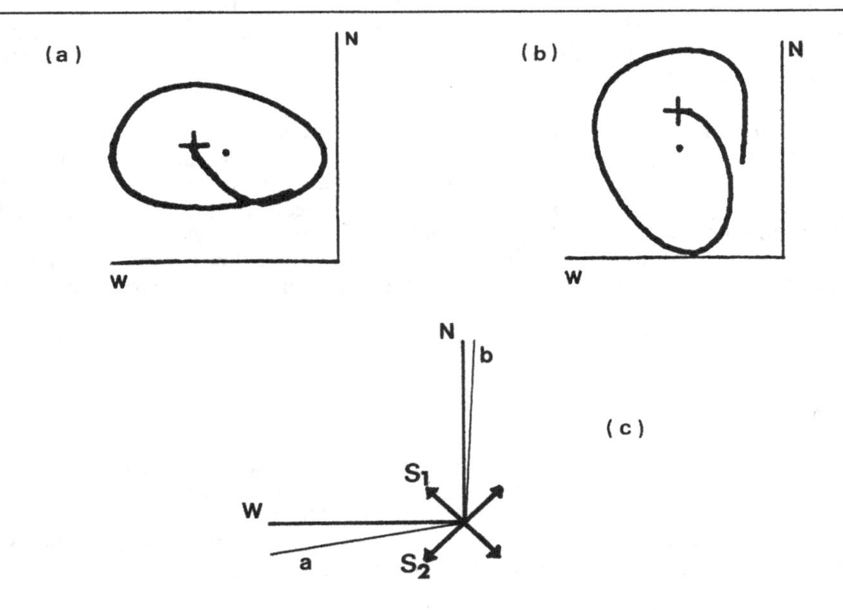

Figure 5-12 Particle motion in the horizontal plane at 1890 m depth for two seismic sources on the surface generating S_H waves polarized N 98° W (a) and N 357° W (b) (Run 1). Principal axes of fast and slow split shear waves (S_1 and S_2) are indicated in (c) (data from the Paris basin, Institut Français du Pétrole, adapted from Stutzman (1990)).

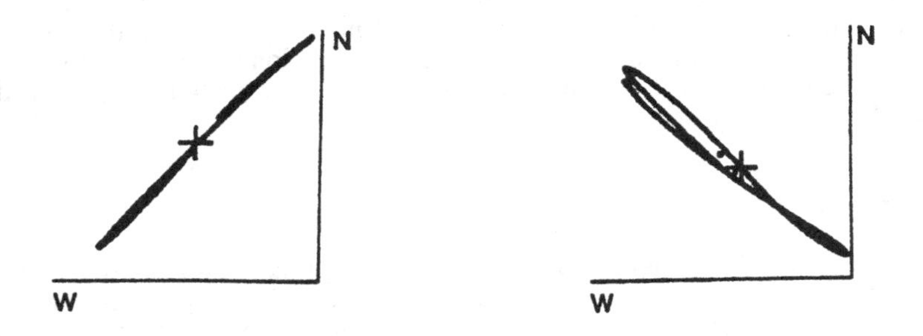

Figure 5-13 The same as in Fig. 5-12 but for Run 2: S_H sources are polarized parallel to the polarization directions of the fast (S_1) and slow (S_2) shear waves (data from the Paris basin, Institut Français du Pétrole, adapted from Stutzman (1990)).

profiling made in a borehole in the Paris Basin by Institut Français du Pétrole. In that experiment two S-wave vibrators were used at the surface as seismic sources and the seismic waves were recorded down to a depth of 1950 m with a 3-component geophone. The seismic signals were emitted at a frequency range of 8 - 50 Hz so that a wavelength in the range 30 - 150 m was observed at the borehole geophone. The first run corresponds to two S-wave sources oriented approximately N-S and E-W. While these sources emitted a quasi-rectilinearly polarized S wave, the horizontal particle motion observed at great depth for each of the sources was of an elliptical character (Fig. 5-12).

The analysis of the particle motion for these two sources leads to a coherent pattern in terms of S-wave splitting, with a fast S_1 wave polarized at N 49° W and a slow S_2 wave polarized at a nearly right angle (N 142° W). The second run corresponds to S-wave sources oriented in directions close to the S_1 and S_2 polarization directions (N 44° W and N 135° W). The result is striking. As expected, at the 1890-m depth the polarization diagrams observed in the horizontal plane show no S-wave splitting, since in each case the S-wave polarization is close to the principal axes of the anisotropic layer lying between the surface and the borehole geophones (Fig. 5-13).

The cross check of this set of data was made by predicting the observation of Run 2 from the data of Run 1 and vice-versa. The fit is excellent so that one can consider an overall anisotropy between the surface and the geophone depth. Similar observation and principal axes directions were found in other wells close to this one (Lefeuvre et al., 1989). The general trend of the stress field in this region exhibits a maximum compression oriented NW - SE. The most satisfactory explanation for this set of observations is thus to invoke a crack-induced anisotropy in this ambient stress field. In a simplified model we can approximate such an anisotropic medium by a transverse isotropy with a horizontal symmetry axis. For a vertical propagation this model predicts fast S waves polarized in a direction perpendicular to the symmetry axis and slow S waves if the polarization direction is parallel to the symmetry axis (Fig. 5-14).

Another evidence for the crack-induced anisotropy in the upper crust is given by Zollo and Bernard (1989). Their paper describes a 3-dimensional particle motion analysis of strong motion records in Imperial Valley, California. It is a careful study of the 15 October 1979, 23:19 Imperial Valley aftershock using high quality acceleration data recorded by 16 stations at hypocentral distances ranging from 12 to 27 km. Two previous studies of that aftershock pointed out the difficulties in modeling the observed accelerograms with the use of a simple velocity structure. Liu and Helmberger (1985) modeled the horizontal acceleration and velocity records by two successive sources of different mechanisms and locations while Gariel et al. (1990) invoked complexities in the rupture time history. Zollo and Bernard (1989) were able to model the observed polarization pattern and, in particular, to explain a significant delay of the East component relative to the North component by considering a simple source model and an anisotropic crust. This anisotropy was modeled under the assumption of a hexagonal symmetry with a horizontal symmetry axis in the upper 4 km of sediments. The model is consistent with the distribution of vertical cracks induced by a regional stress field which is parallel to the cracks.

Brocher (1988) reported on seismic anisotropy observations in the upper 10 km of the continental crust based on refraction experiments. In Alaska, steeply dipping

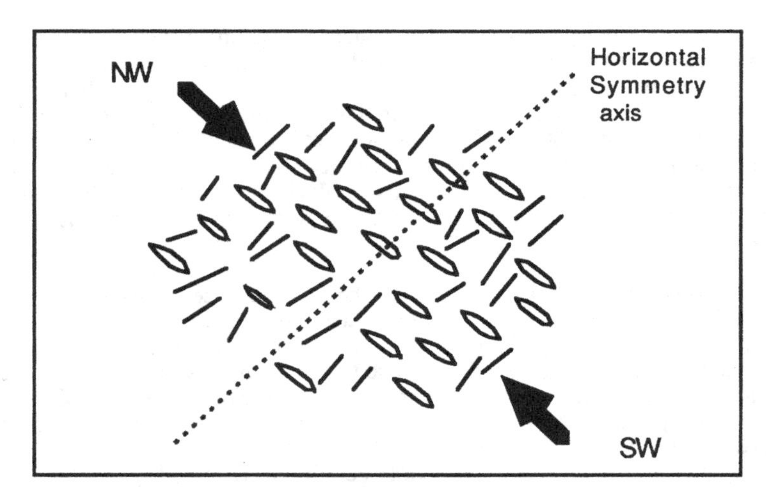

Figure 5-14 Schematic crack model which could explain the shear-wave splitting observed in the Paris Basin from the data displayed in Figs. 5-12 and 5-13. Thick arrows indicate the direction of maximum compression. Pre-existing cracks which are approximately parallel to the direction of maximum compressive horizontal stress may be opened while other cracks are closed. The polarization of fast shear waves propagating vertically is parallel to the direction of preferred orientation of the cracks.

---

phyllites exhibited a 10% P-velocity anisotropy when sampled across and along the foliation planes. In Maine, rocks of low to moderate metamorphic grades also exhibited anisotropy about 10%. The author concluded that the continental structure was extremely heterogeneous but a significant anisotropy could be observed, especially if strata are tectonically deformed into near vertical structures.

Laboratory measurements showed that the average velocity and density of mylonite did not substantially change during deformation under midcrustal conditions from that of protolith (Jones and Nur, 1984). As phyllosilicates have extremely high single-crystal anisotropies (see Chapter 3) and are readily deformed ductilely to form a strong preferred orientation, their fabric will control the seismic anisotropy. The anisotropy of mylonite zones due to the single-crystal anisotropy of phyllosilicates and their preferred orientation can be enhanced by a laminar structure of alternating high- and low-velocity zones. Jones and Nur (1984) estimated their thickness in the case of the Wind River or the Pacific Creek thrust at about 100 - 150 m. With these thicknesses the zones cause reflections in the case of crustal profiling whereas for large wavelengths they behave as a layered anisotropic medium.

Igneous rocks display a velocity anisotropy in the upper crust mainly due to oriented cracks. Coyner (1988) found that Chelmsford granite could be modeled as a transversely isotropic material with the symmetry axis normal to the rift direction. Maupin (1990) brings another evidence for crack-induced anisotropy in the crystalline crust of Scandinavia. The existence of a well developed S_H component of Lg waves excited by explosions in lakes or in the sea, recorded at NORSAR, Norway, can be explained by a mode coupling effect in an anisotropic structure. The anisotropic model which explains the data best contains vertical fluid-filled cracks in the upper 10 km of the crust which are oriented parallel to the regional stress field.

A number of studies confirmed the presence of velocity anisotropy by shear-wave splitting. However, reliable estimates of differential travel times between two orthogonally polarized shear waves are difficult to obtain, mainly because of the waveform contamination by heterogeneities in the crust. Kaneshima and Ando (1988) found evidence for a shear-wave splitting in many seismograms of crustal and subcrustal microearthquakes recorded at a seismic station in the Shikoku Island of Japan. The differential travel times normalized to ray path lengths for subcrustal earthquakes (depth of 40 km) are found to be much smaller than those for crustal events (depths of less than 20 km). The authors thus concluded that the anisotropic region was limited to the upper 10 to 15 km of the crust. This conclusion is drawn from observations of the splitting of steeply propagating shear waves which are sensitive to anisotropy caused by subvertical cracks and faults.

In a detailed study Kaneshima (1990) detected the presence of crustal anisotropy at about 40 seismograph stations in Japan where microseismic shear-wave splitting was documented. At almost all stations, parallel or subparallel alignment of the particle motion direction was observed and the directions of the leading shear-wave polarization are parallel or subparallel to the maximum horizontal compression for the majority of stations though the author found some evident discrepancies. He found at least three phenomena explaining the shear-wave splitting: vertically or subvertically aligned stress-induced microcracks, cracks or fractures located in the vicinity of active faults with the orientation parallel to the fault planes and an intrinsic rock anisotropy resulting from preferred mineral orientations. The travel time differences between leading and slower split shear waves of crustal and upper-mantle earthquakes suggest that the crustal anisotropy caused by nonhorizontally aligned cracks or fractures may be limited to the upper 15 km of the crust with the maximum of fractured anisotropic rock in the uppermost crust (depths of less than 3-5 km).

Gledhill (1990) found evidence for shallow crustal anisotropy in the Wellington Peninsula (New Zealand) located in the top two to three kilometers of the crust, with average delays of shear waves around 0.1 s and a corresponding anisotropy of about 10%. Gaiser and Carrigan (1990) performed vertical seismic profiling (VSP) over a depth interval from 610 to 2690 m and clearly demonstrated that the azimuthal anisotropy of about 5% is confined to depths of less than 600 m.

Very deep borehole seismic experiments may help clarify the depth extent of open cracks which cause the shear-wave splitting. In the cratonic basement in central Sweden, Juhlin (1990) finds VSP evidence of shear-wave splitting, more pronounced (3%) in the uppermost kilometer than deeper in the well (2%). Similarly, Luschen et al. (1991) concluded from P- and S-wave experiments at the KTB-deep drilling site in the Oberpfalz area (SE Germany) that the observed seismic anisotropy in the uppermost 1 or 2 km is probably related to the stress field and the orientation of cracks, whereas between 1.5- and 3-km depth it is mainly caused by the rock foliation. In the Kola deep borehole (Borevsky et al., 1987) instances of heavily fractured granodiorite and mineralized water were found between 4.5- and 9-km depths. Although Leary et al.

(1990) conclude that the microcrack-induced elastic anisotropy may exist throughout the entire brittle crust at least in tectonically active areas, it is probable that a significant portion of this anisotropy is confined to several kilometers of the topmost crust.

Vertical seismic profiling was used in quantitative studies of shear-wave velocity anisotropy. Robertson and Carrigan (1983) recorded orthogonal polarizations of waves generated by a shear-wave vibrator to measure anisotropy in shale, while Majer et al. (1988) and Daley et al. (1988) used similar methods in studies of geothermal reservoir rocks. Although earthquake studies indicated the shear-wave anisotropy already earlier (e.g. Crampin and Booth, 1985; Peacock et al., 1988), the VSP method has been applied to seismically active fault zones only exceptionally. An example is the Cajon Pass deep drillhole in southern California. The project identified a stress-induced anisotropy in basement rocks near the southern San Andreas fault (Li et al., 1988).

Blenkinsop (1990) found strong preferred orientations of fractures and microfractures in oriented cores from the Cajon Pass drill hole in southern California and shear-wave splitting established by VSP experiments. S-wave polarization directions are parallel to the fracture and microfracture orientations, but the anisotropy is caused by the fractures which were probably formed before a late Miocene-Pliocene deformation episode and are not directly related to the modern San Andreas fault or formed by the in situ stress field. Such a conclusion differs from most observations made by other authors, and indicates that the crack geometries must be interpreted with some caution in terms of the in-situ behavior.

A detailed anisotropy study near the San Andreas fault trace at Parkfield was conducted by Daley and McEvilly (1990). The authors found an approximately 8% difference in velocities of S waves polarized parallel and perpendicular to the San Andreas fault. The faster S_H has a particle motion nearly parallel to the fault and the velocity difference seems to decrease with distance from the fault, suggesting that the cause of anisotropy may be the fabric of the fault zone. A similar conclusion that the shear-wave splitting and polarization observations are controlled by the direction of faulting observed at the surface, rather than the direction of the present stress field, was made by Sachpazi and Hirn (1991) from three-component seismograph measurements across the geothermal field of the Milos Island in the Aegean Sea.

A possible physical model with these properties involves a zone of shear fabric near the fault, developed from the continuing slip. Vertical planes of shearing would induce the anisotropy. This simple model cannot explain all the observed particle motions in the wavefield, the data suggest a medium more complex than transverse isotropy with a horizontal axis of symmetry. Blenkinsop (1988) also concluded from the fracture analysis of a fold and thrust belt in Spain that a transverse isotropy, mono-planar or bi-planar crack arrays are inappropriate models of anisotropy. As a matter of fact, in both ancient and contemporary fold and thrust belts, fracture induced seismic anisotropy is more complex due to the presence of multiple fracture sets and folding, and a monoclinic anisotropy is likely to be the general case.

Since cracks as a cause of anisotropy may be sensitive to stress changes occurring prior to an earthquake, it was proposed that temporal changes in S-wave splitting might contribute to the earthquake prediction (e.g., Gupta, 1973a, 1973b; Crampin, 1978; Crampin et al., 1980). This idea was originally related to the dilatancy theory (Nur, 1972; Scholz et al., 1973). This theory which was based on laboratory experiments, described a slight inelastic volume increase and associated cracking occurring in brittle rocks at high stresses prior to failure. The cracking can decrease P- and S-wave velocities and may thus be observed by seismic waves in earthquake source regions.

Brace et al. (1966) observed that stress-induced aligned cracks resulted in velocity anisotropy of rocks, and on the basis of this observation, Gupta (1973a) suggested that dilatancy should have anisotropic characteristics and that this anisotropy might be detectable with shear-wave splitting observations. He also claimed to observe variations in S-wave splitting prior to two earthquakes in Nevada (Gupta, 1973b). However, a later analysis by Ryall and Savage (1974) indicated that the observed variations were not exclusively correlated with earthquake occurrence and that the scattering and phase conversion within complex crustal structures strongly influenced the effects attributed to the S-wave anisotropy. More recent studies (e.g., Zoback et al., 1987) indicated that stresses within crustal rocks were generally far below the yield strength and that dilatancy was probably confined to a very small region near the earthquake source hardly observable by seismic means.

Crampin (1978) and Crampin et al. (1980) developed further the ideas based on the dilatancy theory and suggested that polarization anomalies associated with crack-induced anisotropy might provide a way to monitor dilatancy episodes. The model termed extensive dilatancy anisotropy or EDA (Crampin et al., 1984b) was different from that proposed by Brace (1966) in that it did not require stresses that brake strengths of rocks and did not involve changes in the fluid content of cracks. The model assumed that the crack-induced anisotropy could be a widespread phenomenon and the S-wave splitting observations might permit the monitoring of precursory stress changes prior to earthquakes.

Peacock et al. (1988) observed an approximately 100% increase in S-wave splitting delay times at stations KNW in the Anza network, southern California, and explained it by an increase of stress prior to an earthquake. Later studies by Crampin et al. (1990) reported a decrease in the delay times for this station and led the authors to the conclusion that S-wave splitting might provide means for predicting earthquakes. However, a detailed measurement of S-wave polarizations of local earthquakes recorded at the Anza network by Aster et al. (1990) did not confirm the observations by Crampin et al. (1990) that temporal variations in S-wave splitting delay times at station KNW could be correlated with the occurrence of the North Palm Springs earthquake of July 8, 1986. Aster et al., (1990) thus concluded that while observations of crustal anisotropy have become well established, detailed modelling of the process and its possible use for earthquake prediction remains problematical mainly due to scattering effects and the inability to identify a distinct slow quasi-shear wave pulse. In their reply, Crampin et al. (1991) criticized the automatic procedure used by Aster et al. (1990) for automatic processing of the polarization data and considered that the observed time changes are significant.

Three-component studies of shear-wave splitting support crustal seismic anisotropy with a horizontal symmetry axis as a widespread phenomenon in many different geological and tectonic regimes. The leading split shear wave is polarized in a direction which is usually approximately parallel to the local or regional maximum of horizontal principal stress. The observed anisotropy may have the simple axial symmetry which is explained by effects of fluid-filled subvertical fractures and microcracks, sensitive to a stress field. Anisotropy in the upper crust is not confined to effects of aligned microcracks. In sedimentary basins or in metamorphic regions the crack-induced anisotropy can be combined with the transverse isotropy due to a layering or a schistosity. This superposition of two or more anisotropies leads to orthorhombic or lower symmetry of the anisotropy in the upper crust.

5.4.2 Lower crust

Early models favored basaltic composition for the lower crust of continents similar to this part of the crust beneath oceans. However, recent studies of deep crustal xenoliths and of deep cross sections exposed in orogenic belts display a great variety of petrologies (Fig. 5-15). The lithologies are predominantly mafic and correspond to amphibolite-granulite facies, which may include metasedimentary and metavolcanic rocks, layered mafic complexes, gneisses and ultramafic bodies.

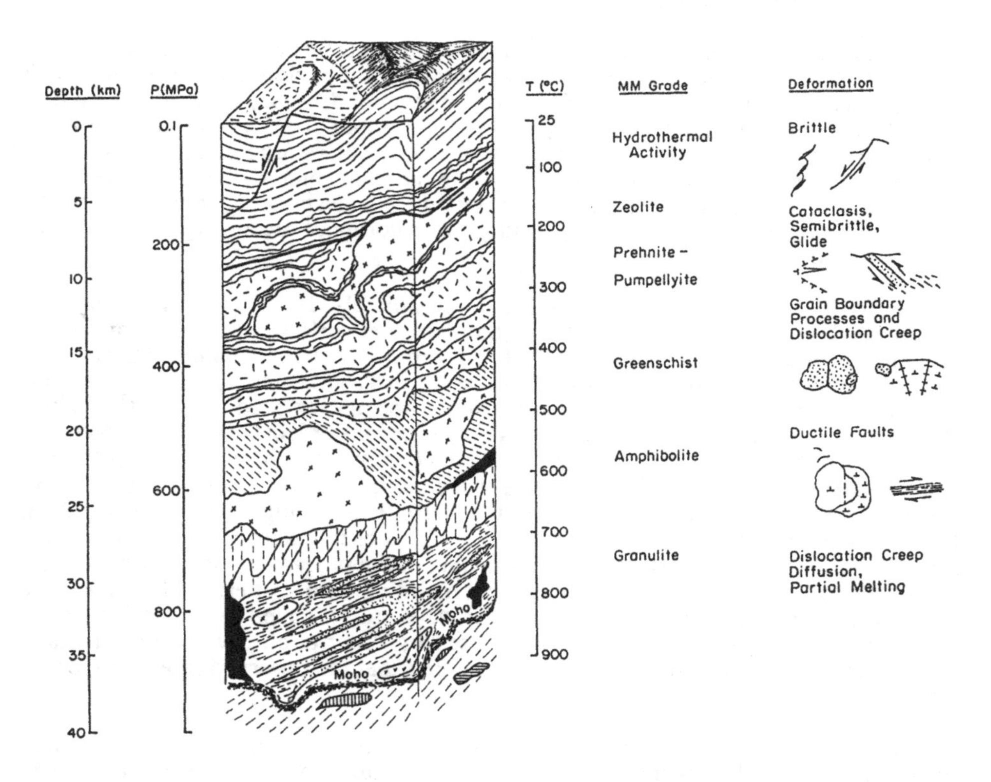

Figure 5-15 Idealized block diagram of continental crustal structure with lithostatic pressure (26 MPa/km), temperature (25°C/km), metamorphic grade and dominant deformation mechanisms (Carter and Tsenn, 1987). (With permission of Elsevier Science Publisher, copyright © 1987).

One of the most distinctive features of the fabric of many rocks comprising the lower crust, is the tendency of their minerals to assume subparallel orientations. The presence of isoclinal folds, boudinage structures and mylonitic shear zones in the deep cross sections attests to the dominance of ductile deformation mechanisms in the lower crust. These structures, coupled with igneous and metamorphic processes, in many cases result in a horizontally layered deep crust from centimeter to kilometer scales (Fountain, 1987). Such a layering in combination with a preferred orientation of minerals in metamorphic rocks produces appreciable seismic anisotropy within the lower continental crust.

The layering of the continental crust was documented by deep crustal reflection profiling which indicated that in many regions the middle and deep crust is pervaded by multiple, discontinuous and flat or gently dipping subparallel reflectors a few to ten kilometers in length. Fig. 5-16 presents an example of a reflection profiling record from eastern Australia (Finlayson et al., 1989) showing a non-reflective upper crustal zone below the sedimentary cover rocks, a reflective lower crust indicating a subhorizontal

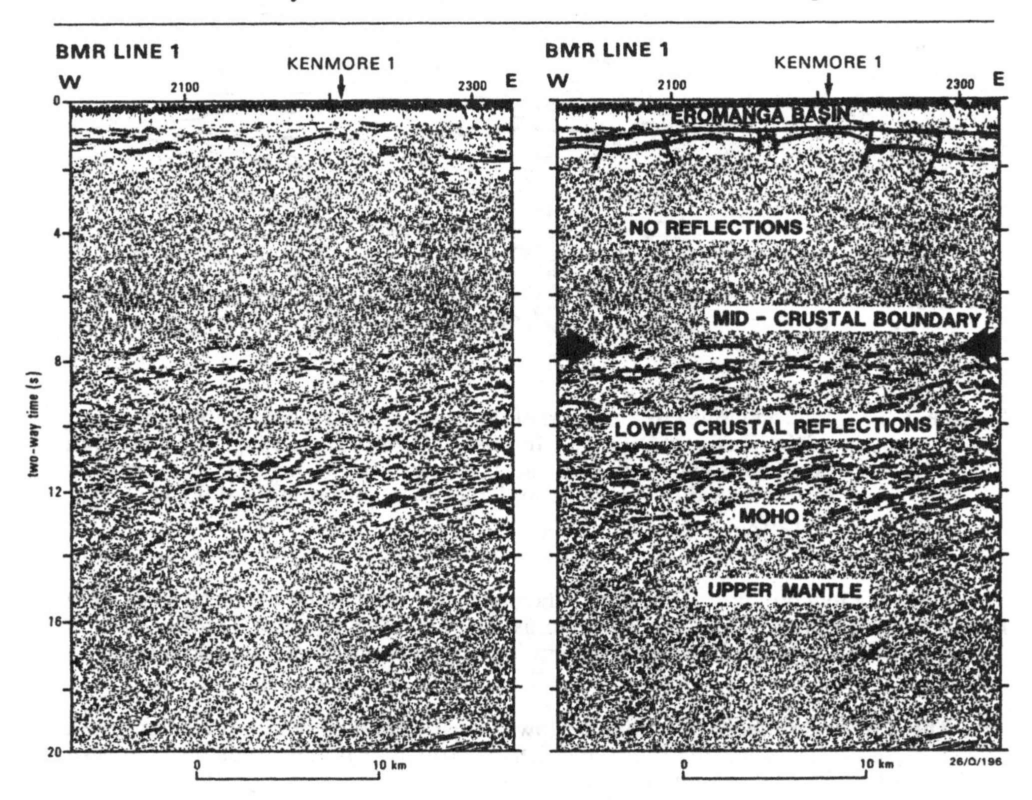

Figure 5-16 A reflection profile from the central Eromange Basin in eastern Australia (Finlayson et al., 1989) illustrating "transparent" upper crust and many reflections in a layered lower crust. (AGU, copyright © 1989).

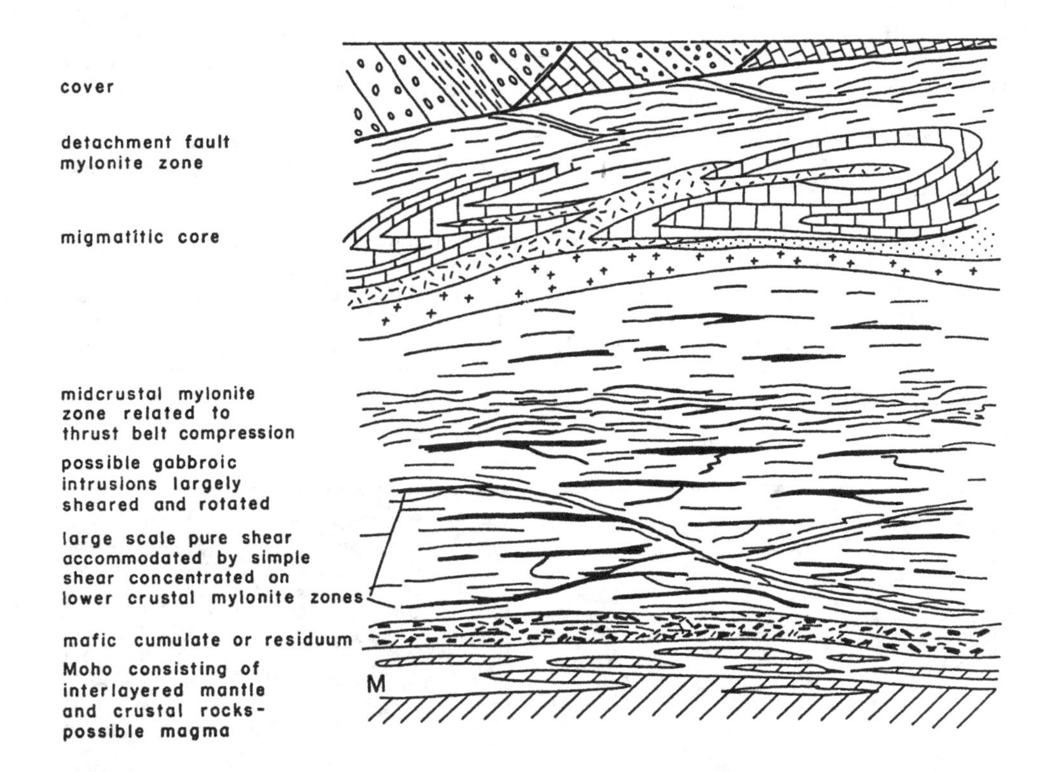

Figure 5-17 Interpretative cross-section of the crust in the northern Ruby Mountains of the Basin and Range (Smithson, 1989). The strong layering, both compositional and structural suggests that most of the crust is characterized by a significant large-scale anisotropy. (AGU, copyright © 1989).

layering and a few dipping reflections in the upper mantle. Similar characteristics were found in many other regions. The layered crust was identified in seismic profiles shot in southwestern Britain, northern and southern France, southern Germany, northern Spain and northern Italy.

The observed reflections within the lower crust were mostly interpreted by strain fabrics, igneous layering, free fluids and thermal effects. All these possible causes of the layering demonstrated by seismic sections would lead to a significant seismic anisotropy. Such anisotropy has not been measured in the field but can be deduced directly from the observed large-scale structures and from laboratory measurements on samples of foliated crystalline rocks (see Chapter 3). The anisotropy probably plays a role also in contributing to reflectivity (Christensen and Szymanski, 1988). Perhaps a typical profile showing a series of anisotropic structures is an interpretative cross-section of the crust of the Basin and Range (Fig. 5-17, Smithson, 1989) based on available reflection data and surface geology.

Up to now, no direct seismological observations exist which would confirm the seismic anisotropy of the lower crust caused by its large-scale layered structure. The wide-angle measurements on three profiles in southwest Germany by Holbrook et al. (1988) show no shear wave splitting on the horizontal component of SmS arrivals, even in the Black Forest, where the finely laminated lower crust would be expected to produce a transversely isotropic medium (Gajewski and Prodehl, 1987). Holbrook et al. (1988) admit an existence of local zones of high seismic anisotropy within the crust, but rule out the possibility of a lower crust in SW Germany with an overall significant anisotropy. On the other hand, near-vertical reflection studies of Luschen et al. (1990) revealed strong S-wave reflections from the lower crust of the same region, which were not observed by previous wide-angle surveys. The authors explain this discrepancy by introducing alternating anisotropic and isotropic lamellae and by a petrological model of deformed amphibolites containing 10-30% of preferentially oriented hornblende. It is obvious that more specially designed experiments are needed to study the seismic anisotropy of the lower crust caused by its large-scale layering.

Several observations indicated azimuthal variations of seismic velocities in the crust. During the deep seismic sounding experiments in the USSR, several crossing profiles showed disagreements in velocities and arrival times which were attributed to both anisotropy and horizontal crustal inhomogeneity. One example demonstrates the Urals region where profiles perpendicular to the mountain range exhibited P velocities

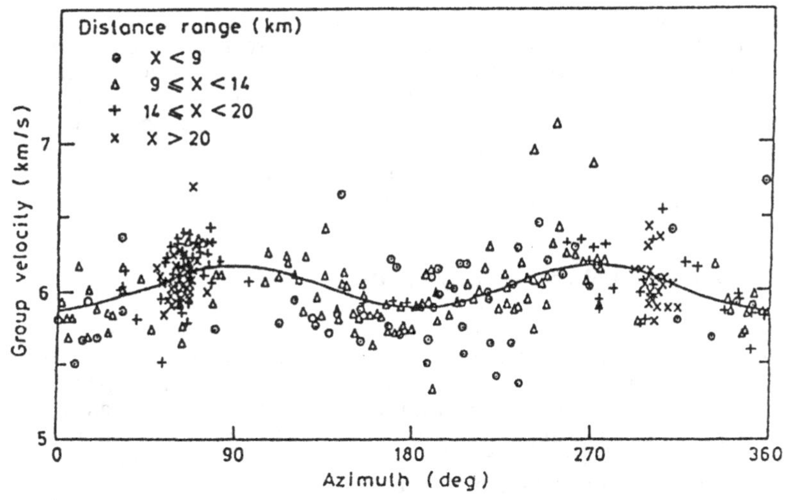

Figure 5-18 Apparent velocity versus azimuth plot in granulite facies rocks of the Arunta Block in central Australia (Greenhalgh et al., 1990). The cyclic pattern testifies to seismic anisotropy of about 5%. The direction of maximum velocity coincides with the predominant direction (strike) of gneissic foliation observed in the outcrop. (AGU, copyright © 1990).

lower than 6.5 km/s, and the profiles orientated along the Urals the dominant velocities between 6.7 and 6.9 km/s (Koshubin et al., 1984). The authors found a more reliable answer to the existence of seismic anisotropy in the crust from S-wave velocities with different polarizations. In the region of the Urals composed of magmatic rocks the anisotropy of S velocities was about 5% and in a 3-4 km thick layer of metamorphic rocks between 7 and 8%. Greenhalgh et al. (1990) determined a 5% P-velocity anisotropy in the granulite facies rocks of the Arunta Block in central Australia (Fig. 5-18), which have been thrust to the surface from lower crustal depths. The velocity maximum is parallel to the strike of the thrusting and to the dominant gneissic foliation observed in an outcrop in the survey region. The anisotropy is explained in terms of preferential alignment of minerals, compositional layering, preferential alignment of fractures, or combination of these factors.

Another example of an interpretation in terms of velocity anisotropy was provided by a closely spaced refraction profiling in a metamorphic complex in southern California (McCarthy et al., 1987). A NE-SW oriented profile, parallel to the extension direction, revealed a mid-crustal layer with P velocity of 6.6 to 6.8 km/s bounded above and below by a laterally discontinuous low-velocity zone (<6.0 km/s). In marked contrast, the NW-SE oriented profile shows a more uniform 6.0 km/s crust down to the crust-mantle boundary. The authors explain the difference in velocities on the crossing profiles either by the 3-D shape of the rock bodies in the middle crust, or by a velocity anisotropy associated with mid-crustal mylonites, which are known to exhibit significant anisotropies. Christensen and Szymanski (1988) determined the P-velocity anisotropy to vary from less than 5% to over 26% at elevated pressures within the Brevard fault zone, which consists of mylonitic metamorphic rocks extending over 700 km from Alabama to Virginia.

Although only a few seismic experiments clearly show the presence of seismic anisotropy in the lower continental crust, many indirect observations indicate its importance in this part of the Earth. Many rocks comprising the lower crust have the tendency of their minerals to assuming subparallel orientations which in combination with many flat-dipping structures and mylonite shear zones, observed in exposed cross-sections, suggest a subhorizontal layering in the deep crust. The layered lower crust, identified also by deep reflection profiling in many regions, can be modelled by a transverse isotropy with the vertical symmetry axis for long wavelengths.

5.5 Subcrustal lithosphere of continents

While a decade ago only a few unambiguous observations of large-scale anisotropy were available in continental regions, at present their number is continuously growing and we can say that the seismic anisotropy is an important property of the deep continental lithosphere. In comparison with the anisotropy in the oceanic upper mantle which reveals a relatively simple pattern, the anisotropy observations in continental regions are more difficult to interpret. This is obviously due to a complex geodynamic development and the resultant pattern of deformations in continents.

The nature of the upper-mantle anisotropy can be examined by studying the velocity anisotropy and mineral orientations within samples of ultramafic rocks which were once part of the upper mantle. Laboratory measurements obtained from multiple samples collected over large areas showed that some ultramafic massifs, which presumably originated within the Earth's mantle, have regional olivine fabric patterns which produce overall anisotropy comparable with observed seismic anisotropy of the uppermost mantle (Christensen, 1984). This was especially well documented for a number of ultramafic bodies within ophiolite massifs (Peselnick and Nicolas, 1978; Christensen and Salisbury, 1979; Christensen, 1984) that represent large slabs of the uppermost mantle tectonically emplaced along faults to the surface.

5.5.1 Observations

First indirect indications of the large-scale continental anisotropy are connected with directional variations of P-wave velocities. Cleary (1967) remarked in connection with a strong azimuthal dependence of the Longshot travel times and the inferred anisotropy beneath the Aleutians that Bolt and Nuttli (1966) and Cleary and Hales (1966) had observed large azimuthal variations in the time terms of Californian stations in the vicinity of the Sierra Nevada arc. Although the azimuthal variations are less pronounced than those of the Longshot, he concluded that the structures underlying island arcs and mountain arcs are of the same kind.

Probably the first irrefutable finding of P-wave anisotropy in the continental subcrustal lithosphere was presented by Bamford (1973; 1977). He realized that in western Germany there existed a number of long refraction profiles with so diverse directions that more or less by accident the overall coverage meets the basic requirement for an objective study of velocity anisotropy - a wide variety of propagation directions through the same material in the upper mantle. Using a time-term analysis of over 80 seismic refraction profiles and fans, the author determined the anisotropy of 7-8% with an azimuth of N 20° E of the maximum velocity attaining 8.3-8.4 km/s. The mean velocity was 8.1 km/s and the total velocity variation exceeded 0.5 km/s.

Another indication of seismic anisotropy in the upper mantle of western Europe was described by Hirn (1977). He associated azimuthal variations of travel times and amplitudes of P waves as well as a possible birefringence of S waves, observed on long refraction profiles with different orientations, with an anisotropic layer near the lithosphere-asthenosphere transition.

While similar studies in northern Britain and the eastern USA did not detect any anisotropy of Pn velocities, this phenomenon was evidenced in the western United States (Bamford et al., 1979). The magnitude of anisotropy was relatively small but significant (around 3%) and a high-velocity direction N 70° E to N 80° E consistent with the offshore refraction measurements of Raitt et al. (1969) for the Pacific upper mantle. The azimuthal dependence of Pn velocities beneath the Transverse Ranges in central California, found by Vetter and Minster (1981), was interpreted by 3-4% anisotropy of the subcrustal material.

High P velocities (8.6-8.9 km/s) interpreted from refraction data for depths between 40 and 100 km (Ansorge et al., 1979; Prodehl, 1981) were sometimes regarded as indirect evidence of velocity anisotropy (Fuchs, 1977; Fuchs and Vinnik, 1982). The Pn velocity of 8.6-8.7 km/s at about a 45-km depth (Ansorge et al., 1979) would produce a 0.03 s^{-1} velocity gradient in southern Germany which exceeds the gradient in a homogeneous geothermally heated medium under self compression by about two orders of magnitude. Without phase transitions with a sufficiently strong

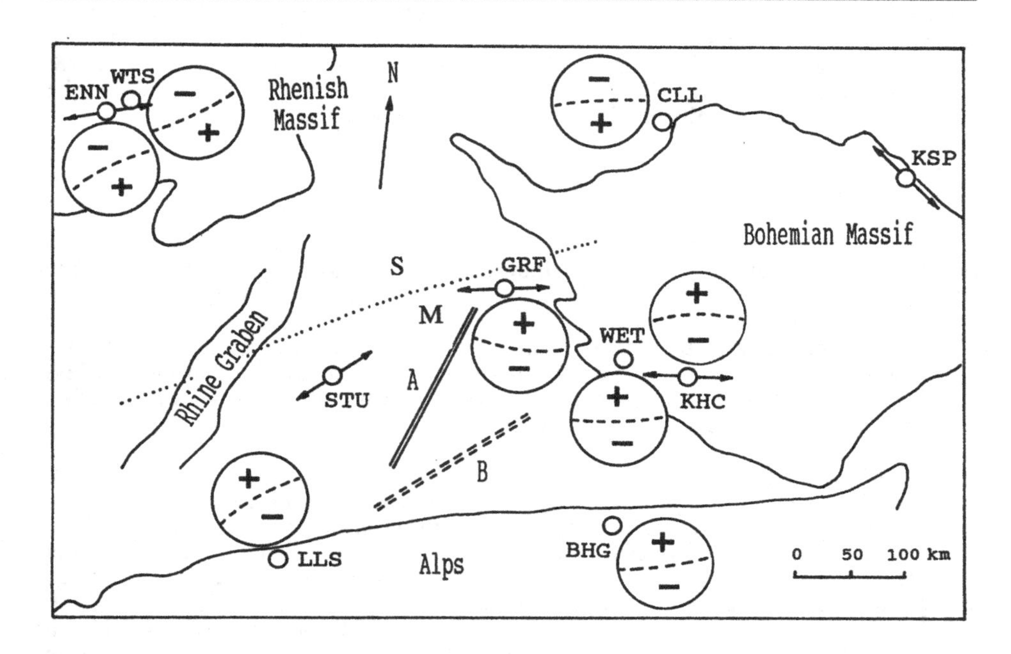

Figure 5-19 Observations of seismic anisotropy in the upper mantle of central Europe: full line A - fast Pn-velocity direction (Bamford, 1977); double dashed line B - fast Rayleigh-wave velocity direction (Yanovskaya et al., 1990); double arrows show polarizations of the fast SKS waves at seismological stations (Kind, et al. 1985; Silver and Chan, 1988; Vinnik et al., 1989a; Makeeva et al., 1990); simplified diagrams of directional terms of relative P residuals (+ for late arrivals, i.e. low velocity directions, - for early arrivals, i.e. high velocity directions; Babuska et al., 1984) suggest opposite dips of the anisotropic structures in the southern part (Moldanubicum -M) and the northern part of the region (Rhenish Massif and Saxothuringicum - S). The suture zone between both parts is dotted. For details see the text.

velocity increase, the high velocities together with the high vertical gradient provide strong indirect evidence of the presence of anisotropy in the subcrustal lithosphere of southern Germany (Fuchs and Vinnik, 1982).

Steep velocity gradients between 30- and 90-km depths, which are incompatible with isotropy, were also derived from long-range controlled source seismic experiments between Ireland and Northern Britain by Bean and Jacob (1990). The authors concluded that the preferred alignment of olivine, which is most probably a product of Mesozoic or early Cenozoic deformation, is an important feature responsible for the seismic anisotropy in northwestern Europe.

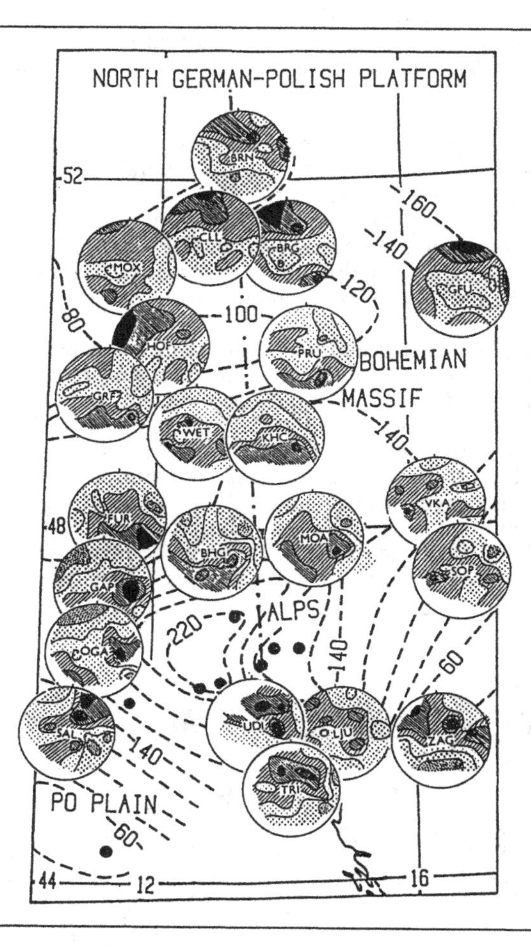

Figure 5-20 Stereographic plots of the azimuth-incidence angle dependent terms of relative P residuals at seismological stations in central Europe portraying directions of relatively high (hatched parts of the diagrams) and low velocities (stippled parts). The diagrams are countoured at a step of 0.4 s and their perimeter corresponds to an incidence angle of 50° at the M-discontinuity, i.e., to that of rays from events at epicentral distances of about 20°. Rays from epicentral distances of 100° are projected close to the centres of the diagrams. Dashed curves are isothickness contours of the lithosphere at 20 km intervals (Babuska and Plomerova, 1989). (AGU, copyright © 1989).

The directional variations of Pn velocities in seismic refraction studies are obtained from the travel times of waves that have travelled more or less horizontally through the uppermost mantle. Thus the information is incomplete, describing variations in a horizontal plane and giving very little indication of velocities in other directions inclined to the horizontal plane, and of the vertical extent of anisotropy as well.

More complete information on P-wave anisotropy might be obtained with the use of seismic waves arriving at observation stations from different azimuths and under different angles of incidence. Such requirements can be met, at least partly, by systematic investigations of the spatial distribution of travel-time curves. Kogan (1981) found a distinct regularity in the distribution of travel-time anomalies over the Earth's surface and distinguished about 40 regions differentiated by travel times of P waves. She concluded that the upper mantle consists of individual blocks, of hundreds to thousands of kilometers in horizontal dimensions and extending to 400-500 km depths which are typical of certain travel times of P waves and the direction of their propagation. If the elastic properties of the upper mantle are anisotropic down to a 200-400 km depth, the effective coefficients of anisotropy are estimated at about 3-7%.

Dziewonski and Anderson (1983) analysed azimuthal variations of teleseismic P arrivals. They refined P-wave travel times and determined station corrections for almost 1000 seismological stations. The corrections, which exhibit general consistency over broad geographic areas, involved three terms; the static effect and two cosine terms. While the static and the first azimuthal terms are connected with Earth's inhomogeneities, the authors assume that the second azimuthal term is due to upper mantle anisotropy. They found that the slow directions determined from the second azimuthal terms correlated with the maximum stress direction in the crust.

Babuska et al. (1984) analysed directional terms of the relative residuals of P waves approaching seismological stations from different regions in order to get information about spatial variations of P velocities in the deep lithosphere in dependence of azimuths and incidence angles of arriving waves. The differences in the relative residuals are systematic and large, between 1 and 2.5 s at individual stations, and thus their cause has to be sought mainly in the mantle. Stations with similar patterns of the diagrams form groups which extend over several hundred kilometers (Fig. 5-20). The diagrams have a more or less consistent pattern of negative residuals for waves arriving at a station from one side and positive ones for waves from the opposite side. The pattern of the diagrams changes at significant tectonic boundaries, i.e., in central Europe

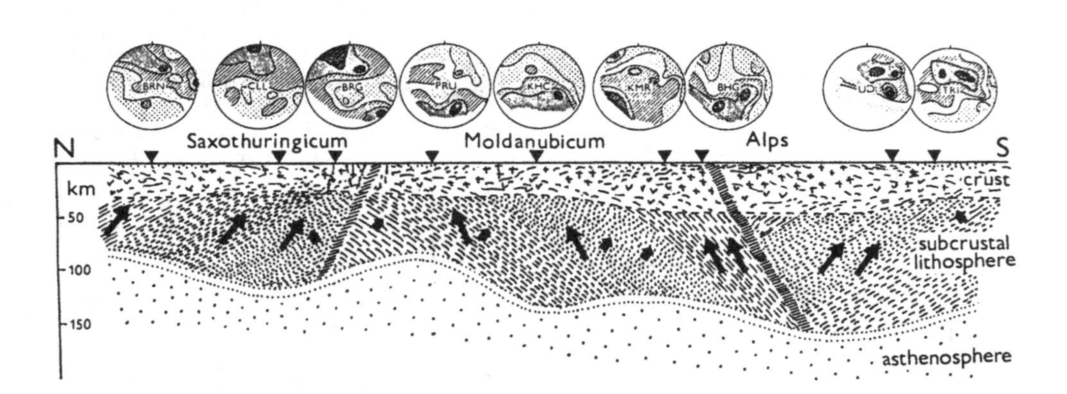

Figure 5-21 Cross-section showing a model of oriented anisotropic structures in the subcrustal lithosphere of central Europe along the profile between stations Berlin (BRN) and Trieste (TRI). Sectors of negative residuals are hatched, sectors of positive residuals are stippled. Variable inclined hatchings in the subcrustal lithosphere schematically represent systems of dipping anisotropic structures. The long and short arrows show directions of relatively high and low velocities, respectively, of P waves arriving from azimuths close to the direction of the cross-section (Babuska and Plomerova, 1989). (AGU, copyright © 1989).

at the suture zone between the Saxothuringicum and the Moldanubicum (Fig. 5-21). An analysis of more recent data (Babuska and Plomerova, 1991) confirmed the previous directional dependence of the relative residuals which was interpreted as large dipping anisotropic structures in the subcrustal lithosphere. The large-scale P-velocity anisotropy of these structures (6-11%) agrees with the small-scale anisotropy established by laboratory measurements on olivine-rich ultramafites. In the case of dipping structures, the anisotropic component of spatial variation of P residuals exhibits 2π periodicity (Babuska et al., 1987) in contrast to the results of 2-dimensional azimuthal analyses of P residuals, where only π terms are related to anisotropic structures (Dziewonski and Anderson, 1983).

It has been often pointed out that observations of the directional dependence of P velocities are affected by lateral inhomogeneities, because, in fact, the waves travel in different media. As pointed out by Todd et al. (1973), in-situ observations of acoustic double refraction, or shear-wave splitting, are more favorable in this respect. Problems regarding lateral heterogeneities are reduced because the pair of shear waves travels along the same propagation path. Furthermore, the difference between the split shear waves is related to the magnitude and symmetry of the anisotropy and thus a series of observations are potentially powerful for determining the in situ anisotropy.

Records of teleseismic shear waves from permanent seismological stations and temporary arrays furnish important information about anisotropic structures. For probing the deep lithosphere, the SKS wave which travels through the mantle as an S wave and the outer core as a P wave, is often used to search for the azimuth of the fast polarization direction and the delay time between split arrivals for each phase. The polarization of the SKS phase is independent of source properties and in an isotropic Earth this wave should be polarized strictly as an S_V wave. If long-period SKS propagate in an anisotropic medium, this linear particle motion becomes elliptic and the search for the anisotropy is done by removing the energy of the transverse component. The method was described by Vinnik et al. (1984) and then used in many regions.

Kind et al. (1985) gave evidence for the presence of seismic anisotropy in the lithosphere beneath southern Germany and concluded that the continental anisotropy was a relatively small-scale phenomenon compared with the oceanic anisotropy. Silver and Chan (1988) found azimuthal anisotropy localized in the top 200 km of the upper mantle at nine stations in North America and Europe.

Makeeva et al. (1990) determined shear-wave splitting in SKS phases at two stations in the Bohemian Massif. Approximately E-W directions of polarization of the fast S wave are consistent with similar results for western Europe. The value of the time delay between the two split shear waves is close to 1 s implying that the thickness of the anisotropic zone in the upper mantle exceeds 80 km. The typical time delays around 1 s were also found by Ansel and Nataf (1989) beneath 7 stations of the Geoscope broadband network. The directions of polarizations of the fast S wave at 3 continental stations in America could be related to plate tectonic processes.

Vinnik et al. (1989a) analyzed observations of low-frequency SKS waves recorded on horizontal components at several stations of the GEOSCOPE and NARS arrays. They noticed several discrepancies between different types of anisotropy observations, like between the dominant E-W direction of fast polarizations in North America from SKS data and the dominant N-S fast velocity direction of long-period

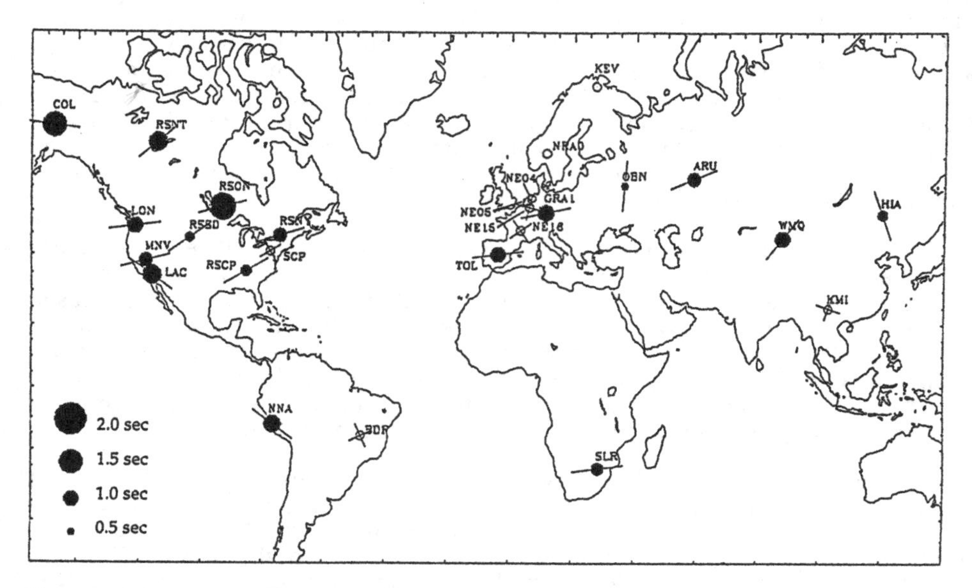

Figure 5-22 World map of SKS splitting results (Silver and Chan, 1991). Solid symbols give the delay time between arrivals of both split S waves and the lines show orientations of the fast S wave. Open symbols denote the absence of detectable splitting from at least two non-orthogonal back azimuths. (AGU, copyright © 1991).

surface waves (Montagner and Tanimoto, 1990), the fast polarization direction N 140° E at station SSB in the Massif Central, which differs from E-W directions generally observed in western and central Europe.

Shear-wave splitting on both SKS and direct S phases at temporary and permanent stations in western Nevada and southern California was observed by Savage et al. (1990). For stations in the northern Basin and Range, the results were consistent yielding an average of fast polarization azimuths N 75±8° E and an average time delay 0.9±0.3 s. For the northern stations, the anisotropy appears to be a "fossil" phenomenon associated with pre-Miocene extension. For the California station the fast polarization is nearly parallel to the strike of the San Andreas Fault and the authors assume that it is related to the shear strain associated with the relative plate motion.

Shear-wave splitting of SKS phases for China was investigated by Chan and Silver (1990) from digitally recorded seismic data from five stations situated both in stable and tectonically active regions. The results provide an opportunity to interpret the relationships between geological processes and the mantle strain in a conterminous area with contrasting geological features.

In 1990 the observations of SKS splitting existed nearly at 100 stations in many regions (Vinnik, 1990, personal communication), yielding typical delay times of both subvertically propagating shear waves around 0.8-1 s. The fact the shear-wave splitting was observed at a great majority of the stations clearly indicated that the large-scale seismic anisotropy is a wide spread property of the deep continental lithosphere.

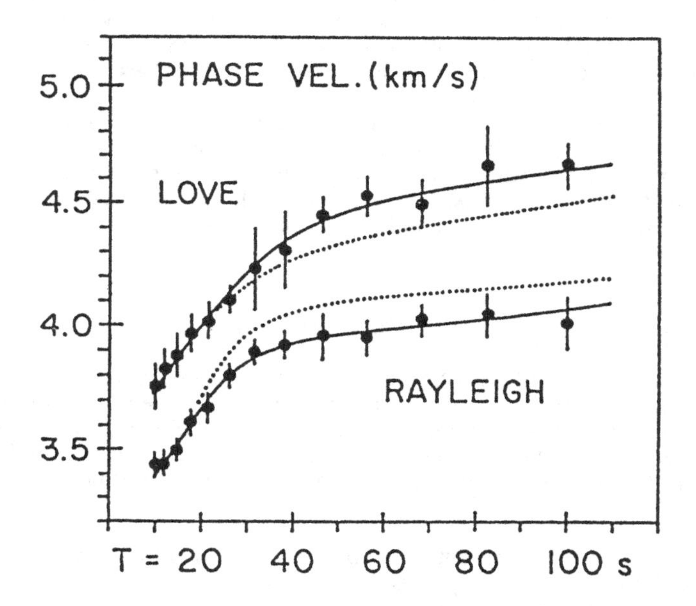

Figure 5-23 Observed (symbols) and modeled phase velocities (Wielandt et al., 1987). The incompatibility is explained by the presence of an anisotropy in the upper mantle beneath the Bohemian Massif.

Many indications of anisotropic properties of the continental lithosphere come from surface-wave observations, mainly from Love/Rayleigh-wave incompatibility (see sections 2.5 and 4.5). The earlier observations of incompatibility between Love- and Rayleigh-wave dispersion curves were observed in the western United States by McEvilly (1964) (see Fig. 4-12). Other possibilities for seismic anisotropy in the continental lithosphere have been suggested by different authors after anomalies were observed in the Rayleigh-wave particle motions. Crampin and King (1977) for example found anisotropic polarization in higher-mode surface waves across Eurasia implying about a 10-km thick anisotropic layer beneath the M discontinuity, and Neuhofer and Malischewsky (1981) observed an anomalous polarization of Love waves along paths in Eurasia and explained this observation by anisotropy in the Earth. However, as we have seen in section 4.5 these observations have to be considered with great caution since dipping interfaces can cause similar anomalies in an isotropic model.

Wielandt et al. (1987) found apparent incompatibility between the Rayleigh-wave and Love-wave data recorded at broadband seismological stations KHC and KSP in central Europe (Fig. 5-23). Maupin and Cara (1991) found a similar discrepancy in the Iberian Peninsula. In all these surface-wave observations, Love waves appear as abnormally too fast when compared with theoretical curves computed from models fitting the Rayleigh-wave observations. Wielandt et al (1987) have inverted both Love- and Rayleigh-wave phase-velocity curves into a model of the shear-wave velocity versus depth and they have found a significant discrepancy between the resulting shear-velocity models in a depth range from the M discontinuity down to about 217 km. To get a smooth velocity-depth distribution which satisfies the dispersion data, the authors had to introduce an anisotropy of the shear-wave velocities.

Yanovskaya et al. (1990) determined azimuthal anisotropy of Rayleigh waves in central and western Europe with the anisotropy coefficient 2.6% for a period T=25 s and 1.4% for T=50 s. The azimuths of maximum velocity range between 40° and 60°, which means that the velocity maximum is oriented close to the strike of the Variscan belt in central Europe. However, the authors admitted that the azimuths were not well constrained by the data.

5.5.2 Petrological constraints for interpretations

The observed effects of anisotropic wave propagation described in the previous section are so large that they cannot be explained by crustal effects, and their cause has to be sought in the upper mantle. Direct evidence of the petrology of the upper mantle is derived from xenoliths in volcanic eruptions, from ophiolites and ultramafic massifs, which were uplifted by tectonic processes from the mantle along faults, especially in collision zones.

The most important representatives of the upper mantle materials as to their physical properties are peridotites, pyroxenites and eclogites. A recent analysis of heavy mineral concentrates from upper mantle samples made by Schulze (1989) confirmed the previous estimates of an average composition of upper-mantle xenoliths and led to the conclusion that the upper mantle was predominantly peridotite. Moreover, the author found that the amount of eclogite in the upper 200 km of the subcontinental mantle was perhaps less than 1% by volume overall. As the velocity anisotropy of both eclogites and pyroxenites is substantially smaller than the anisotropy of peridotite (see Chapter 3), it is reasonable to base any model of seismic anisotropy of the subcrustal continental lithosphere on the elastic properties of peridotite and lherzolite specimens (Maaloe and Steel, 1980).

Oehm et al. (1983) analyzed upper-mantle xenoliths from basalts of the Northern Hessian Depression and came to the conclusion that the average modal composition of 73 vol.% olivine, 18% orthopyroxene, 7% clinopyroxene and 1.3% spinel was close to xenoliths from the Eifel, Massif Central and Spain, and from worldwide sampling (Wedepohl, 1987). It is thus reasonable to take such a composition as a basis for a model of the subcrustal lithosphere beneath continents. It is interesting to note that an average composition of the oceanic uppermost mantle is very similar to analyses made for continents. Dick (1987) made modal analyses of 266 peridotite samples dredged at 6 ocean ridges and gave an average composition of 74.8 ± 5.3% olivine, 20.6 ± 3.7% enstatite, 3.6 ± 2% diopside, 0.5 ± 0.2% Cr-spinel, and 0.5 ± 1.2% plagioclase.

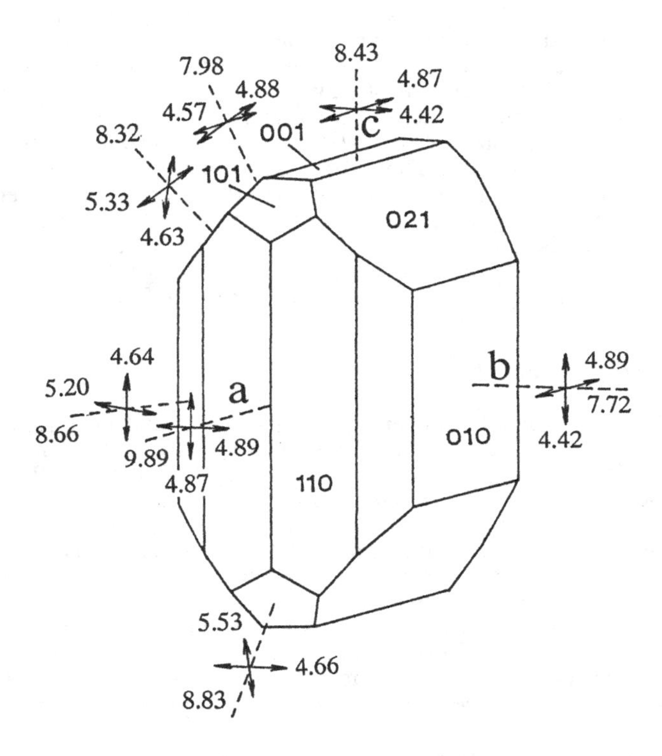

Figure 5-24 Compressional and shear-wave velocities in a monocrystal of olivine. Original data from Kumazawa and Anderson (1969). (AGU, copyright © 1969).

Olivine, orthopyroxene (enstatite or bronzite, both with similar elastic properties) and diopside have single-crystal anisotropies, which could contribute to the overall anisotropy of their aggregates. Christensen (1984) showed, however, that an admixture of pyroxene decreased the anisotropy of olivinic rocks. Their average P-velocity anisotropies vary between 7.5 and 10.8% whereas in pyroxenite and eclogite they only attain 3.5 and 2.3%, respectively (Babuska and Plomerova, 1991).

Shear-wave velocities and their anisotropy have to be evaluated with particular caution because their values in anisotropic media depend on both the directions of wave propagation and the polarizations of shear waves. In general, the larger number of directions and polarizations is used for investigating a massive rock, the larger variations of velocities and their anisotropy are observed. This may be one of the reasons why the S-wave anisotropies in the same rock type, like in dunite from Twin

Sisters, vary between 4% (Simmons, 1964) and 9% (Babuska, 1972b).

A special case of velocity anisotropy measurements are studies of shear-wave splitting in a single direction of wave propagation. This applies to e.g. studies of large-scale anisotropy by means of SKS polarization analysis, which, however, represent only subvertical directions of wave propagations. Since in anisotropic media the velocities vary with the directions of wave propagation relative to the orientation of symmetry elements, the estimates of the anisotropy determined by this method (e.g. Vinnik et al., 1989a; Silver and Chan, 1988) may not always be representative as the measurements are only exceptionally orientated in the directions of extremal velocities. As an example we can imagine the shear-wave splitting for wave propagations and wave displacements along the crystallographic axes of a single crystal of olivine (Kumazawa and Anderson, 1969). The anisotropy of S waves propagating along b- and c-axes attains 10% whereas along a-axis it is practically zero (Fig. 5-24).

5.5.3 Anisotropic models

As in the deep oceanic lithosphere, also in the continental subcrustal lithosphere the preferred orientation of olivine crystals is generally accepted as the principal cause of the observed seismic anisotropy. In order to fit both the observations of Pn anisotropy and the velocity-depth distribution derived from long refraction profiles, Fuchs (1983) proposed an "anvil" model of seismic anisotropy in the upper mantle beneath southern Germany. The model is relatively complicated as it introduces two different olivine orientations, however, both have the crystallographic axes either horizontal or vertical. The author deduced from the orientation of a stress field in the crust that the preferred orientation of olivine in the topmost mantle is achieved by a leakage of the crustal stress field into the mantle. It would mean that the formation of the olivine orientation is a recent process which could be related to recent mantle flow in the Rhine graben. Most of the other authors, however, interpret the mantle anisotropy as a "fossil" feature.

As we mentioned in paragraph 5.5.1, the long-range seismic profiles between Ireland and Northern Britain revealed three zones with steep velocity gradients at depths between 30 and 90 km. Both the pattern and velocity of the arrivals are incompatible with isotropy and, therefore, Bean and Jacob (1990) suggested a model consisting of three horizontal anisotropic layers embedded in an otherwise isotropic lithospheric mantle consisting of 80% olivine and 20% orthopyroxene. The layers with widths of 8, 8 and 3 km have their centers at 45, 70 and 85 km depths with the maximum P-wave anisotropy of about 6%. The authors found no apparent correlation between the orientation of anisotropy and the maximum horizontal compressive stress direction in the region and they concluded that olivine alignments were thus likely to be relic "frozen in" features. Contrary to the "anvil" model of Fuchs (1983), they argue that the upper mantle is not necessarily a strain marker for the last major orogenic episode since it may undergo deformation which decouples from the brittle upper crust and hence is not "transmitted" to the Earth's surface. Bean and Jacob (1990) explain the anisotropic layers by large-scale shearing zones, whose origin is difficult to imagine, though. Moreover, generally a much thicker anisotropic layer is needed to account for the observed anisotropic effects.

Assuming a transverse isotropy with a horizontal symmetry axis and an effective S-velocity anisotropy of about 4%, which corresponds to average or slightly lower than average anisotropies of olivinic ultramafites (see Fig. 5-25), Silver and Chan (1988) estimated that each second of the time delay between the fast and the slow S waves corresponded to an anisotropic layer 115-km thick. Allowing 0.2 s delay time for crustal anisotropy would reduce these estimates by about 25 km. The authors interpret the shear-wave splitting as due to "fossil" anisotropy localized in the top 200 km of the upper mantle. Later they found a good correlation between delay time and lithosphere

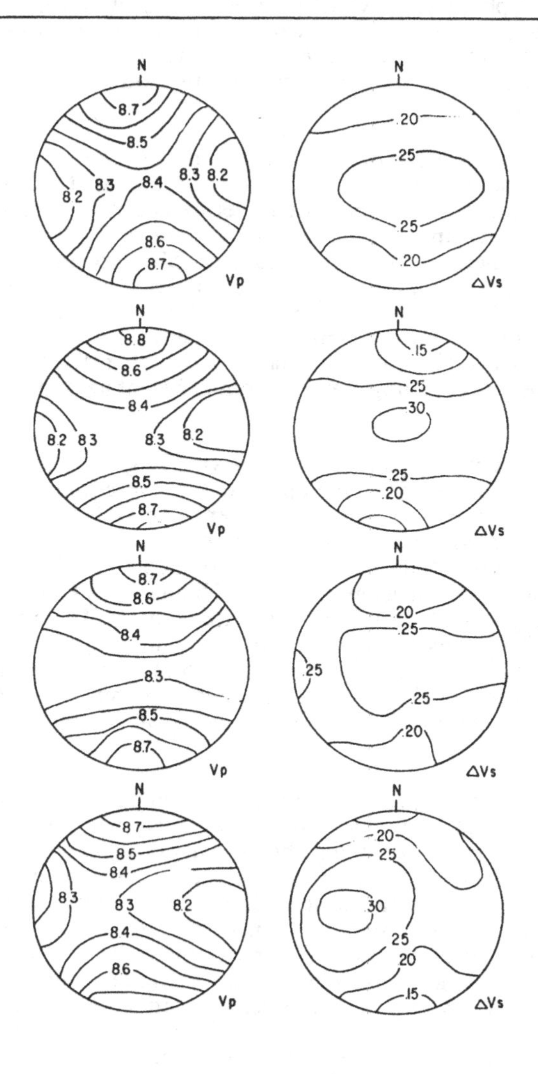

Figure 5-25 Anisotropy of compressional wave velocity and shear-wave splitting in km/s computed by Christensen (1984) from petrofabric data for olivine and pyroxene. From the top are ultramafic rocks of ophiolite massifs in Twin Sisters (USA); Red Mountain, New Zealand; Dun Mountain, New Zealand; Red Hills, New Zealand. (With permission of the Royal Astronomical Society, copyright © 1984).

thickness (Silver and Chan, 1991) which supports several previous suggestions that a substantial amount of the observed seismic anisotropy is localized in the subcrustal continental lithosphere.

For the Bohemian Massif, Makeeva et al. (1990) estimated the anisotropic layer to be thicker than 80 km beneath stations KHC and KSP. This is in accord with thicknesses of the subcrustal lithosphere which were determined beneath these stations at about 100 km (Babuska et al., 1987), and which probably represent a principal source of this anisotropy. A 100-km thick mantle layer characterized by 4% S-velocity anisotropy is also estimated by Savage et al. (1990) beneath the southern Basin and Range province and the Mojave Desert.

In the papers dealing with interpretations of seismological data from the viewpoint of anisotropy, mostly hexagonal and orthorhombic anisotropies are treated either with horizontal or vertical axes of symmetry. Such simplified interpretations are influenced by the generally subhorizontal propagation of surface waves and of P waves in the controlled-source refraction experiments, as well as by subvertically propagating SKS phases, whose fast velocity direction and a time delay between two S waves are only determined in the horizontal plane; the authors then speak about "azimuthal" anisotropy. However, vertical and horizontal orientations of symmetry axes are not consistent with spatial variations of P residuals at seismological stations, which reflect relative compressional velocities of waves approaching stations from different azimuths under different angles of incidence.

Consistent patterns of negative residuals for waves arriving at groups of stations from one side and positive ones for waves from the opposite side were interpreted by large dipping anisotropic structures in the subcrustal lithosphere (Fig. 5-26). In a search for a model which could explain the different observations of seismic anisotropy in central Europe, Babuska et al. (1991) select on the basis of papers reviewing the anisotropy of upper mantle rocks (e.g. Christensen, 1984; Nicolas and Christensen, 1987) a "typical" peridotite aggregate with preferably orientated olivine. The aggregate has the P velocities 8.7, 7.9 and 8.15 km/s along orientated a-, b- and c-axes of olivine, respectively, and it is made of 3 components: olivine with the orthorhombic preferred orientation, olivine with the hexagonal preferred orientation and a mixture of randomly oriented olivine, orthopyroxene and clinopyroxene crystals yielding an isotropic material with V_P=8.28 km/s.

The elastic constants of this model aggregate were used to compute an orientation which is compatible with the observations of anisotropy in central Europe derived from SKS splitting, spatial variations of P residuals and the azimuthal dependence of Pn waves. By rotating the model aggregate and checking various orientations of symmetry axes, Babuska et al. (1991) find that the first two observations derived from teleseismic body waves were compatible with the model of inclined anisotropic structures, whereas the azimuthal variations of Pn velocities needed a different orientation of the model aggregate. The authors explain this inconsistency by the variable orientation of symmetry axes in different depth horizons of the subcrustal continental lithosphere. While the SKS splitting and the spatial variations of P residuals reflect an averaged anisotropy within the whole lithosphere thickness, the Pn velocity variations give information only about the top part of the subcrustal lithosphere. Thus the Pn observations by Bamford in southern Germany can only be explained under an assumption of vertically changing olivine orientations as suggested by Ishikawa (1984)

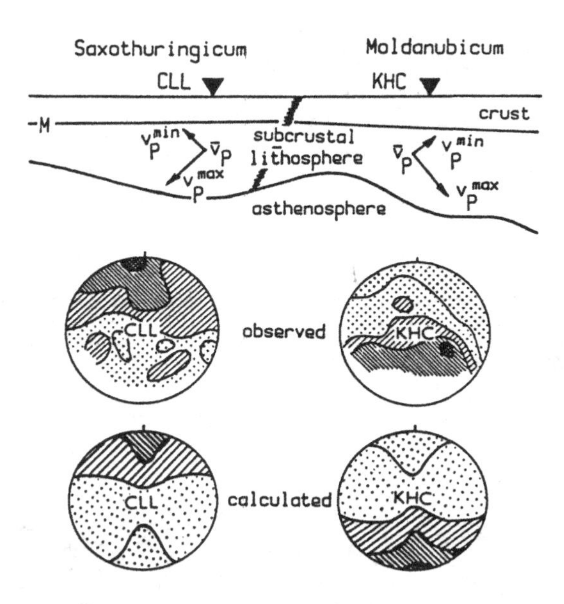

Figure 5-26 Comparison of P-residual diagrams contructed from observed data at seismological stations KHC and CLL in the Bohemian Massif and the diagrams calculated for a model of inclined anisotropy in the subcrustal lithosphere. In this model of orthorhombic symmetry the velocity maximum V_P^{max} = 8.7 km/s is oriented along the dip of the structures, the velocity minimum V_P^{min}= 7.9 km/s perpendicular to the structures, and the intermediate velocity $\bar{V}_P$ = 8.2 km/s points along their strike; the anisotropy coefficient equals 9.5% (Babuska and Plomerova, 1989). For explanation of the diagrams see Figs. 5-20 and 5-21. (AGU, copyright © 1989).

for the oceanic subcrustal lithosphere (see Chapter 5.2). Lateral changes of the olivine orientations might also be an important phenomenon within the deep continental lithosphere.

One of the main targets of the investigations into seismic anisotropy is to obtain information on strain and large-scale tectonics in the upper mantle. In this respect, an important question arises in connection with the origin and age of the observed anisotropies caused by olivine preferred orientations.

As we already mentioned, most interpretations assume that the seismic anisotropy in the deep continental lithosphere is a "fossil" feature. Estey and Douglas (1986) concluded that the lithosphere would have a "fossil" alignment of olivine and pyroxene which occurred when the material was at a temperature higher than 1100-1200° K and

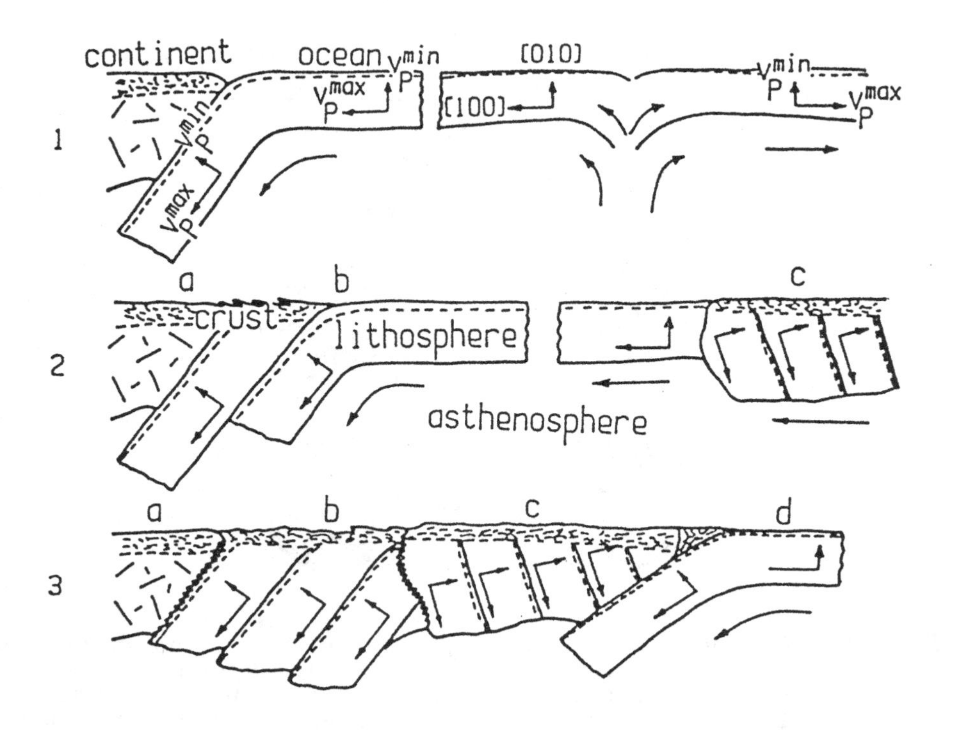

Figure 5-27 Three stages of a continental growth by a succession of accretions in a series of subductions which produce systems of large anisotropic structures in the deep continental lithosphere (Babuska and Plomerova, 1989). The letters denote an old continent (a), an accreted terrane (b) by which the continent was enlarged as a result of successive subduction processes, a later accreted exotic continental fragment (c) with a different orientation of deep structures, and a series of new subductions of the oceanic lithosphere (d). (AGU, copyright © 1989).

the underlying asthenosphere should have actively maintained orientations of these minerals related to the in-situ mantle flow.

Silver and Chan (1988) interpreted the splitting of SKS phases at the stations in Archean greenstone terrains of the Canadian Shield by 2.5-2.7 billion years-old anisotropic fabric which was due to a series of deformational episodes at the end of the Archean. The azimuth of fast polarization directions tends to be parallel to the regional trend of mountain belts, both in the Archean of the Canadian Shield (Silver and Chan, 1991) and in the Hercynian system in central and western Europe (Vinnik et al., 1989a).

Vauchez and Nicolas (1991) stressed the importance of tectonic movements parallel to the strike of orogenic belts and suture zones during the continental collision and oblique subduction, and suggested that the anisotropy could be best explained by the mantle flow during orogenesis parallel to the mountain belt. According to the authors, the major strike-slip faults are 5-20 km wide in the upper crust and in the subcrustal lithosphere the regional extent of reoriented peridotites may be several tens of kilometers wide perpendicular to the fault. However, the seismological data indicate that the large-scale anisotropy is probably a general property of the subcrustal lithosphere of continents found at many places at large distances from major strike-slip faults. The model by Vauchez and Nicolas (1991) also implies a subhorizontal orientation of olivine a-axes and thus is not compatible with inclined velocity extremes derived from the spatial distribution of relative P residuals, which were interpreted by systems of paleosubductions of an ancient oceanic lithosphere (Babuska et al., 1984).

The proposed "mega-domino" model of successive paleosubductions (Babuska et al., 1991) is characterized by relatively high velocities (a- and c-olivine axes) oriented within the plane of subduction and by low velocities (b-axes) approximately perpendicular to that plane (Fig. 5-27). The basic assumption that the olivine orientations are preserved in a substantial part of the subducted lithosphere, is quite realistic with regard to the rigidity, large volume and relatively low temperature in the subducted plate. Under such conditions a preferred crystal orientation once formed is not easy to destroy (McKenzie,1979). Moreover, several observations suggest that the seismic anisotropy exists in the recently subducting plates or above them (e.g. Ando et al., 1983). McKenzie (1979) showed that the high strain induced by subduction above the sinking slabs should produce important anisotropy with olivine a-axes oriented in the dip direction of the sinking slab. Such olivine orientation may enhance the anisotropy in a part of the subduction zone.

The proposed model is thus characterized by an inclined "layering" of anisotropic structures. The seismic anisotropy within the individual layers can vary due to both a variable amount of oriented olivine crystals and variable orientation of a- and c-axes in the plane of subduction. The variability of the anisotropic structures within the continental lithosphere is further enhanced by other phenomena, such as a destruction of smaller blocks in collision zones, or by a reorientation of olivine crystals under changing physical conditions (e.g. rifting, deep strike-slip faults).

Equivalents of the systems of inclined paleosubductions within the deep continental lithosphere, proposed by Babuska et al. (1984), can also be seen in observations of other authors. Cook (1986) interpreted dipping structures in the deep crust of accreted terranes from seismic reflection data. His model of continental "shingling" (Fig. 5-28) within the upper lithosphere has a pattern similar to the systems of paleosubductions. The inclined reflectors (e.g. McGeary and Warner, 1985; Warner and McGeary, 1987) which penetrate from the crust into the mantle may represent mega-thrust zones at the contact of two subsequent subductions.

Mooney et al. (1987) interpreted seven hundred kilometers of seismic refraction profiles from the accretionary Chudach terrane in southern Alaska and concluded that the sequence of continuous, north-dipping alternating low- and high-velocity layers were a stack of subducted oceanic plates. The deepest plate is the youngest and currently subducting plate while the upper plates have already been incorporated into the

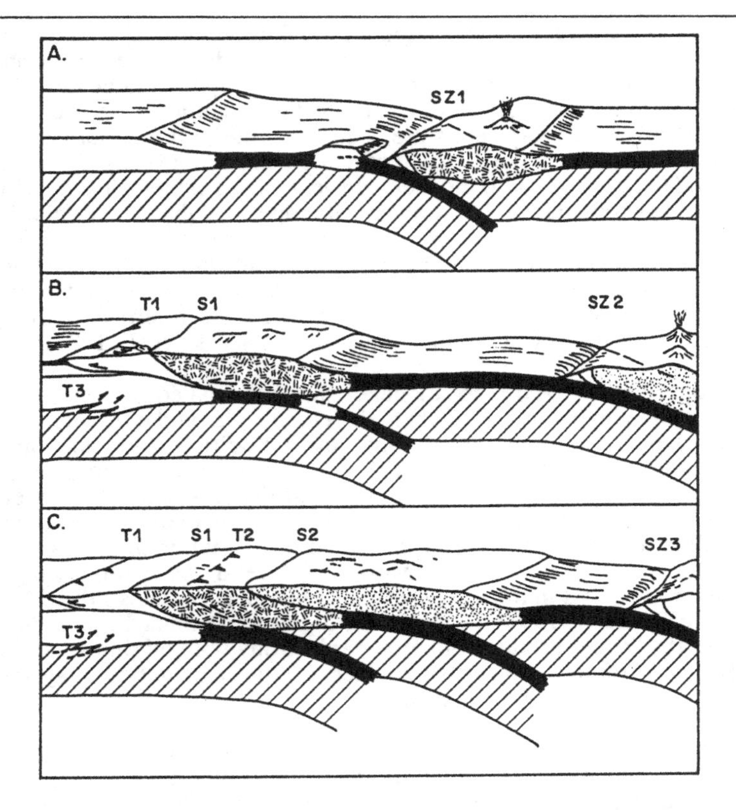

Figure 5-28 Model of continental shingling proposed by Cook (1986). In A, the subduction of the ocean lithosphere (SZ1) beneath an arc produces some features such as seamounts, to be decapitated. Partial subduction of continental lithosphere (B) results in the traping of allochtons above the continental margin and the development of a low angle suture (S1). Cessation of subduction results in the creation of new subduction zones outboard of the enlarged continent (C). (AGU, copyright © 1986).

continental crust. This interpretation implies that the continental growth in southern Alaska is in part accomplished by the underplating of paleosubduction zones at the active continental margin. On the basis of similar results obtained on the Vancouver Island to the South, the authors conclude that underplating may be a generally important mechanism in the growth of the continental crust.

The model of inclined anisotropic paleosubductions of the ancient oceanic lithosphere "frozen" in the deep continental lithosphere is supported by a number of geological and geochemical data, which lead us to the conclusion that various parts of continents originated in oceanic settings through plate-tectonic processes, such as subduction and accretion of the oceanic lithosphere at convergent boundaries, or, island arcs and continental collisions. For example, Roden et al. (1990) measured Sr and Nd isotopic compositions and rare earth element abundances in inclusions from the Navajo Volcanic Field of the Colorado Plateau and concluded that the continental lithosphere largely appears to be accreted oceanic lithosphere to depths at least as great as the spinel-garnet peridotite transition. It is thus highly probable that additions to continents from oceanic sources provided large volumes of the oceanic lithosphere rigid enough to retain the preferred olivine orientations in the deep continental lithosphere.

In comparison with the oceanic subcrustal lithosphere, where the seismic anisotropy is determined by subhorizontally oriented high P velocities and subvertical low velocities, the anisotropy pattern within the deep continental lithosphere is more complicated, being characterized by "mega-domino" systems of inclined anisotropic structures (Fig. 5-29). Such a model showing a basic difference in anisotropies of oceans and continents derived from body-wave observations, is supported by the recent global upper mantle tomography of Rayleigh- and Love-wave velocities and anisotropies made by Montagner and Tanimoto (1990). At shallow depths (100 km) a large dissymmetry is observed between oceans and continents. The transverse isotropy with a vertical symmetry axis is larger (around 3-4%) beneath oceans where $S_H>S_V$. On the other hand, beneath continents the transverse isotropy is smaller than 2% and in some parts $S_V>S_H$. This ratio may depend on dips of anisotropic structures within the subcrustal continental lithosphere (Fig. 5-29).

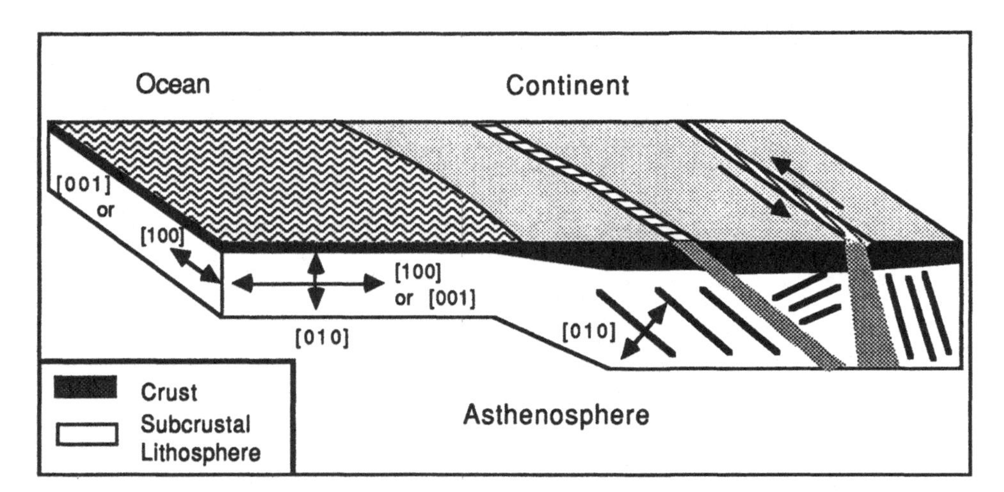

Figure 5-29 Schematic cartoon portraying orientations of anisotropic structures in the oceanic and continental subcrustal lithospheres. Thick bars show directions of high P velocities ((010) plane parallel to [100] and [001] olivine axes).

The existence of large-scale anisotropy in the deep continental lithosphere has been demonstrated by independent seismological observations in many regions. It is thus a great challenge for the near future to search for anisotropic models which would fit different observations of seismic anisotropy and to use such models for geotectonic interpretations.

Chapter 6 - Anisotropy in deeper parts of the Earth

The oceanic lithosphere exhibits a clear seismic anisotropy. In continents, there are many observations which strongly suggest that seismic anisotropy is present in large enough volumes of the upper crust and subcrustal lithosphere to make direct seismic investigations possible. In deeper parts of the Earth, evidence of seismic anisotropy is more difficult to obtain. At sublithospheric depths, hypothesis testing thus plays an important role in interpretations of seismic anomalies. Interpretations of the seismic anomalies which are suspected to be due to anisotropy in the asthenosphere are commonly conducted in the framework of preferred orientations of upper mantle minerals. Due to its large abundance, to its strong elastic anisotropy and its ability to be preferentially oriented in a shearing deformation zone, olivine plays a major role in modeling the asthenospheric anisotropy in connection with the present flow pattern of the upper mantle convection. Effects of other mineralogic constituents, such as orthopyroxene and clinopyroxene in the upper mantle, and higher pressure phases and/or constituents with different chemistry at greater depths have to be taken into account (Anderson, 1989). Finally, in the deepest part of the mantle, it is not impossible that region D" at the core-mantle boundary is also anisotropic. Several recent studies concerning both normal mode data and inner core PKIKP body waves have led different authors to suggest that the inner core might also be anisotropic. Seismic anisotropy of the liquid outer core is very unlikely but, even there, it has been suggested that a well organized pattern of convection in the liquid core with a rain of solid particles in cold regions could induce heterogeneities in the seismic velocity pattern which might show up as an apparent seismic anisotropy! Seismic anisotropy has thus been invoked at all depths as a possible explanation of observed seismic anomalies.

Below the lithosphere, the most numerous studies of seismic anisotropy concern the asthenosphere. Therefore, in this chapter we start by reviewing our present knowledge of anisotropic properties of the asthenospheric layers. Results and speculations concerning the upper-mantle transition region below 400 km depth, the D" zone near the core-mantle boundary and the inner core, are reviewed in the second section.

6.1 Asthenosphere

The concept of the asthenosphere does not come from seismology but from several investigations concerning isostatic compensation over geological time scales and from the observation of the post-glacial isostatic adjustment at a shorter time scale (Barrel, 1914; Daly, 1940). A low-viscosity layer beneath a rigid elastic lithosphere is required in order to explain why isostasy exists only on a regional scale and not on a local scale, and why a perturbation of the isostatic equilibrium due to an overload on the surface (ice sheet or thick deposit of sediments) eventually cancels out as time goes on. The suggestion by Gutenberg that a zone of low P-wave velocities was likely to exist at sublithospheric depths (e.g. Gutenberg, 1948, 1959), the suggestion by Caloi (1954) that a shear-wave phase observed on seismograms at teleseismic distances could be guided within this "asthenospheric" low-velocity channel (he called the corresponding seismic waves "S_a wave") and, finally, the confirmation that a broad shear-wave low-velocity zone at depths between 100 and 250 km was necessary to explain the shape of the dispersion curves of surface waves in several parts of the world (Takeuchi et al., 1959; Dorman et al., 1960), led geophysicists to think that the asthenosphere and the so-called low-velocity zone (LVZ) were very likely to share a common origin. The high temperature existing in the upper mantle became the most likely candidate to explain both the decrease of seismic velocities and the viscosity at depth.

A second type of seismic observations, the attenuation properties of seismic waves, yielded additional constraints on the physical processes acting in the asthenosphere. Interpreted in terms of unelastic processes, surface-wave data have shown that the intrinsic attenuation increases drastically at depth when passing from the lithosphere to the asthenosphere (Anderson and Archambeau, 1964). As the geotherm is probably very close to the melting point of the upper mantle materials in the asthenosphere (e.g. Anderson and Sammis, 1970), interpretations of both the LVZ and the highly attenuating zone in terms of partial melting were quite commonly accepted in the sixties (Anderson, 1966). Waff (1980), while studying the possible distribution of the liquid in the upper mantle partial melts, showed that partial melting was unlikely to cause a large enough reduction of seismic velocities and an increase of attenuation to explain the seismic properties of the LVZ. He favored other mechanisms of attenuation such as a grain boundary relaxation (Goetze, 1969) or a dislocation-impurity interaction (Gueguen and Mercier, 1973). Dislocation climb and dislocation glide in olivine crystals were proposed to provide a more plausible explanation for both the viscosity and the seismic properties of the asthenosphere (e.g. Minster and Anderson, 1980).

The boundary between the lithosphere and the asthenosphere is not defined on a unique basis. Anderson (1989) recognizes five types of lithosphere: flexural, "kinematic", chemical, thermal, and finally seismic. In principle, both the flexural and the seismic lithospheres should refer to the same physical property: elasticity. But, even there, incompatibility occurs between the flexural and seismic thicknesses of the lithosphere since they are viewed on different time scales. The thickness of the flexural lithosphere is usually defined as the thickness of the elastic plate which can account for the shape of the bent lithosphere near subduction zones or its depression under the load of volcanic islands (e.g. Watts et al., 1980). This process takes several hundred thousand to several million years to take place so that the elastic properties which are appealed to correspond to long-term static deformations. The seismic lithosphere, in turn, is defined as corresponding to the limit between the upper high-velocity lid and the seismic LVZ.

The LVZ has been studied from several types of seismological data. Teleseismic P-wave residuals are sometimes used to put constraints on the lithospheric thickness (e.g. Chapter 5). But the most convincing evidence comes from shear waves, either by using synthetic seismograms of S_H body waves (e.g. Grand and Helmberger, 1984) or surface waves. To determine the lithosphere thickness, the depth-variations of the upper mantle shear-wave velocities are often represented by a two-layer model, one layer for the high-velocity lithospheric lid and one layer for the LVZ. The elastic properties elicited by seismic waves correspond to very short time constants, of the order of minutes for long-period surface waves. The time constant involved in the definition of the flexural thickness of the lithosphere is larger by more than ten orders of magnitude.

The elastic rheology of mantle minerals is thus very likely to depend on the duration of the applied stresses, in particular, if such a large time span is considered (e.g. Minster and Anderson, 1980). Not surprisingly, smaller lithospheric thicknesses are obtained when we consider the long-term flexural properties (flexural or, improperly, "elastic" thickness) than when we look at the seismic velocities (seismic thickness). Attempts to reconcile the flexural and seismic lithosphere thickness were made using a concept of a sharp and time-independent transition between the lithosphere and the asthenosphere (Regan and Anderson, 1984; Estey and Douglas, 1986; Anderson, 1989). However, as we will see, more recent seismological investigations suggest that a smooth transition between the high seismic velocity lid in the lithosphere and the asthenospheric LVZ might be a better parameterization of the seismic velocity distribution with depth (Nishimura and Forsyth, 1989), suggesting that the lithosphere-asthenosphere boundary is likely to correspond to thermally activated and

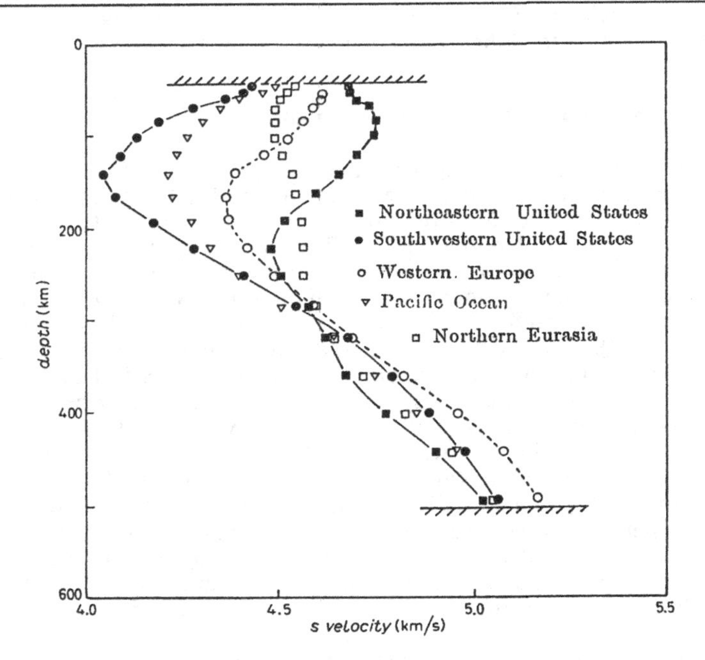

Figure 6-1 Shear-wave velocity models obtained in the framework of isotropic inversions in different tectonic provinces. The data used to build up these models are Rayleigh-wave dispersion curves, including overtones obtained from stacking techniques on wide aperture arrays of WWSSN stations (after Cara, 1983). A pronounced low-velocity zone is apparent in most investigated regions, except beneath northern Eurasia. The LVZ is more pronounced and shallower in young geological provinces (western U.S. and the Pacific) than in old continents (northeastern U.S.).

time-dependent physical properties of the upper mantle materials.

All the seismic models of the asthenosphere obtained before 1980 were built up within the framework of an isotropic upper mantle. When smooth depth-velocity profiles are considered, such as in Fig. 6-1, the definition of the seismic thickness of the lithosphere can be made in a fairly flexible way, for example by assuming that a given value of the shear-wave velocity is linked to an isotherm, then connecting the seismic lithospheric thickness with the thermal thickness. As we have mentioned above, Anderson and Regan (1983) have raised the important question of the validity of such a seismic model if the Earth's upper mantle is anisotropic, thus reconsidering the discussion on the discrepancies which are observed between the flexural and seismic lithospheric thicknesses. This shows us that anisotropy can be at the center of discussions which are apparently rather far away from the question of seismic anisotropy itself. This emphasizes again the importance of the problem of trade-off between anisotropy and heterogeneity. We will come back to this question later on but let us first describe which different steps have led many seismologists to consider the asthenosphere, as well as the lithosphere, to be very likely anisotropic.

6.1.1 The depth extent of large-scale anisotropy

One of the earliest pieces of evidence of anisotropy in the asthenosphere came from the observations of incompatibility between the long-period Love- and Rayleigh-wave dispersion curves related to the structure of the Earth beneath the Pacific Ocean (Forsyth, 1975; Schlue and Knopoff, 1976). The first interpretations of the observed Love/Rayleigh-wave discrepancy were made in terms of a shape-induced anisotropy due to a thin layering of the upper mantle with partial melts trapped in horizontal thin layers or penny-shape pockets preferentially oriented in the horizontal plane (Schlue and Knopoff, 1976, 1977). If partial melts were concentrated in such thin horizontal layers, the asthenosphere should display an overall anisotropy with a vertical axis of symmetry. The medium should then be transversely isotropic and could be described by five elastic parameters (see Chapter 2). Horizontally propagating SH waves should be faster than the horizontally propagating S_V waves, and Love waves should appear as being abnormally fast compared with Rayleigh waves, as frequently observed. If such an explanation were valid it might have important consequences on the flow pattern of the upper mantle beneath the lithosphere since strong viscosity anisotropy would then be expected in such a layered medium (e.g. Honda, 1986). On the other hand, the possibility of an anisotropic viscosity in the upper mantle is a major problem for modeling geoid signal and postglacial rebound observations (Christensen, 1987).

In fact, even if it is clear that partial melting occurs in the asthenosphere, it is not necessary to invoke a shape-induced anisotropy with a thin horizontal layering of the medium, or penny-shape pockets of partial melts, to explain the observed incompatibility between the Love- and Rayleigh-wave dispersion curves. Even if there are physical reasons to believe that the geometry of partial melts in the gravity field may be organized in a horizontally layered structure in regions of stable upper mantle conditions, such as beneath large oceanic basins or stable continental regions (Waff, 1980), the preferred orientation of olivine crystals provides an alternative explanation which is based on more plausible hypotheses than the *ad hoc* model of partial melts. Olivine crystals represent at least 60 % of the upper mantle materials and thus form the major constituent of the upper mantle. Olivine crystals can easily be oriented in an overall shear strain. This mechanism was advocated as a likely cause of deep seismic anisotropy in the pioneering work of Forsyth (1975), who was the first to observe simultaneously an incompatibility between the Love- and Rayleigh-wave dispersion curves at a long period and an azimuthal variation of the Rayleigh-wave velocity in the Pacific. Forsyth (1975), however, did not make any conclusion on the depth extent of the anisotropy, the asthenospheric anisotropy appearing more as a possibility than a proved fact in his first study. Schlue and Knopoff (1976) used longer period waves and were able to present more convincing arguments showing that the asthenosphere was very likely to be anisotropic.

Before discussing in greater detail the question of a possible origin of the asthenospheric anisotropy, let us first address the basic question of its existence and depth extent. Most of the information on this subject is known from surface-wave data, essentially from phase and group velocities of long-period Love- and Rayleigh-wave fundamental modes. Unfortunately, the depth resolution of such data is quite poor as far as the anisotropic parameters are concerned. The reality of the asthenospheric anisotropy became questionable, for example, after two sets of fundamental mode data related to the Pacific were interpreted in opposite ways in the late seventies, either by considering an anisotropic asthenosphere (Schlue and Knopoff, 1977) or an anisotropic lithosphere (Yu and Mitchell, 1979). One of the main reasons for this controversy is the limited depth resolution of fundamental-mode data. As a much better resolution is obtained when using higher-mode surface waves, we will report here first on the results obtained from higher-mode data.

As mentioned in section 4.5, stacking techniques applied to large aperture arrays of long-period stations are necessary to isolate the different branches of higher mode surface waves and to perform accurate-phase velocity measurements. In a series of papers, Cara (1978, 1979), Cara et al. (1980), Leveque and Cara (1983, 1985) and Cara and Leveque (1988) have applied different stacking techniques to extract the higher-mode dispersion curves of Rayleigh and Love waves in North America, the Pacific, and western Europe. The phase velocities of the first modes of Love waves measured across the whole Pacific are displayed in Fig. 4-13 together with the curves computed for a model fitting the Rayleigh-wave data in the same area. Similar curves for western Europe and North America are shown in Fig. 6-2 and Fig. 6-3, respectively. In the three cases, Love-wave velocities appear to be larger than those predicted from isotropic models fitting the Rayleigh-wave dispersion curves over the same region. This is particularly clear in the Pacific and western Europe. As the Rayleigh-wave models used in these computations were obtained by inverting both fundamental-mode and higher-mode data in the period range 30-100 s, the depth resolution was quite good and there is no doubt that the observed Love/Rayleigh-wave incompatibility is real and is not due to abnormally low velocity in some unresolved parts of the Rayleigh-wave model.

The data of Cara et al. (1980) for western Europe are less reliable than those obtained in the Pacific because they were obtained from a single event. They were only inverted in terms of isotropic shear velocity models with the use of a separate inversion scheme for Love and Rayleigh waves. Although such a procedure is not correct from a strictly theoretical point of view as we have seen in section 4.5, it gives us at least the order of magnitudes for the seismic anisotropy which could make the data compatible. Inferences made on this basis show that differences between the shear-wave velocities obtained separately from Love- and Rayleigh-wave data are larger than 0.1 km/s and

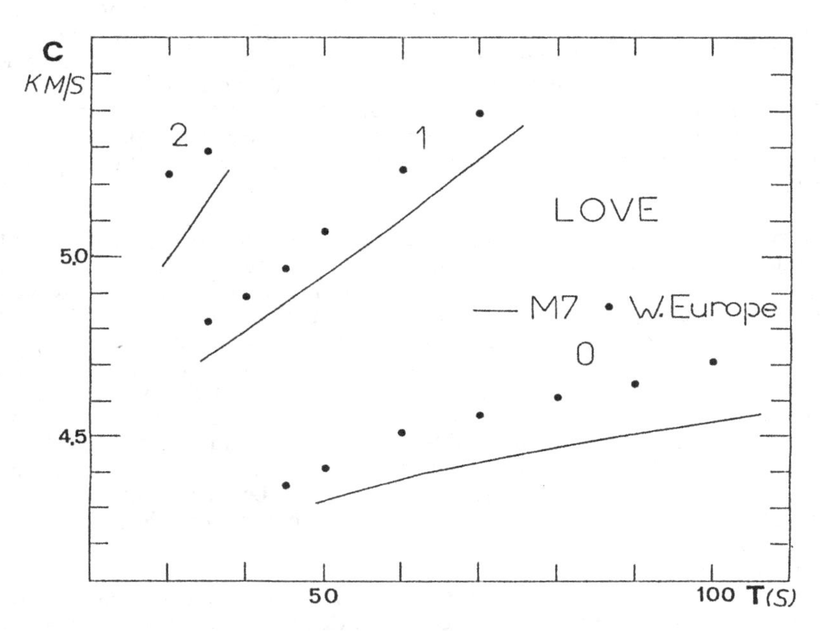

Figure 6-2 Average Love-wave phase velocity C versus period T in western Europe from the North of Norway to the South of Spain. Theoretical curves computed for model 7 of Nolet (1977), based on Rayleigh-wave data, are shown for comparison (modified from Cara et al., 1980). The observed velocities are too fast for both the fundamental mode (mode 0) and the two first higher modes (modes 1 and 2). (With permission of the Royal Astronomical Society, copyright © 1980).

may extend to a 150-km depth. A more recent study by Dost (1990) based on the data from the NARS array across south-western Europe has provided similar dispersion curves for the fundamental mode of Love waves and its first higher mode at the long period but not for the second mode which is very poorly fitted by model A1 of Cara et al. (1980) anyhow (see Fig. 6-2). But Dost (1990) did not succeed in extending these Love-wave dispersion curves toward shorter periods, probably because of the very strong lateral heterogeneity pattern of the crust in this region, which could alter the significance of the data due to diffraction and multipathing effects. Dost's (1990) dispersion curves are related to a shorter array of stations extending between Denmark and the South of Spain, while Cara's et al. (1980) dispersion curves are related to a broader area extending to the North of Norway. Thus the right significance of these curves is not clear. One can just notice that they are not in agreement with horizontally averaged isotropic models derived from Rayleigh waves in this complex area made of different tectonic provinces and that, again, Love waves appear as being too fast up to the periods which are very sensitive to the elastic properties of the asthenosphere.

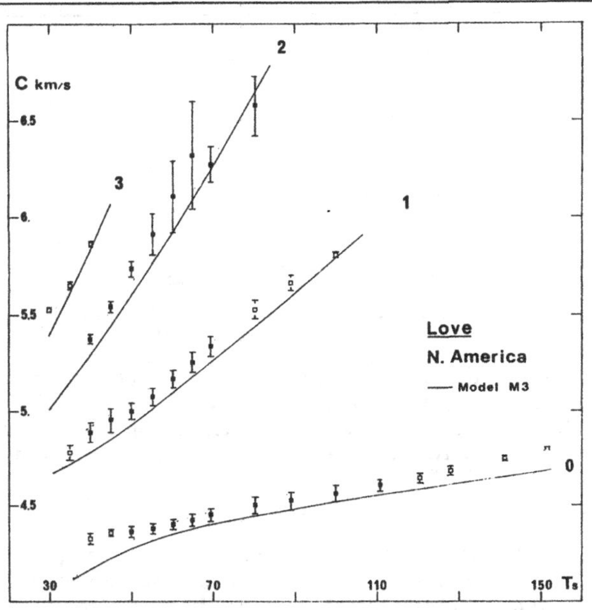

Figure 6-3 Average Love-wave phase velocity C, versus period T, in North America. Theoretical curves computed for model M3 of Cara (1979) based on Rayleigh-wave data are shown for comparison (after Leveque and Cara, 1983). The observed velocities are slightly too fast for the fundamental mode (mode 0) and the first higher modes (modes 1-3). (With permission of Elsevier Science Publisher, copyright © 1983).

Better quality higher-mode data were obtained for North America and the Pacific. Eight events were used by Cara (1978) to retrieve the Rayleigh-wave dispersion curves in these two regions and two events were used by Leveque and Cara (1983, 1988) for the Love waves. These data were first interpreted in terms of a transversely isotropic model of the mantle by assuming no other *a priori* constraints than the symmetry around the vertical axis. The results of the inversion of the two data sets are shown in Fig. 6-4 for the parameter ξ which is the square of the ratio between S_H and S_V velocities. As expected from the sign of the Love/Rayleigh-wave discrepancy, S_H waves are faster than S_V waves in both cases. These differences seem to exist to at least a 200-km depth in the Pacific and a 400-km depth beneath North America. Evidently, anisotropy is resolved at depths larger than the lithospheric thickness in both regions. In these two regions, the asthenosphere thus appears as an anisotropic layer exhibiting a few per cent shear-wave velocity anisotropy.

Concerning the problem of trade-off between the lithosphere thickness and anisotropy which was pointed out by Regan and Anderson (1984), let us note that this is a general question in seismic estimates of the lithosphere thickness. Using arrival

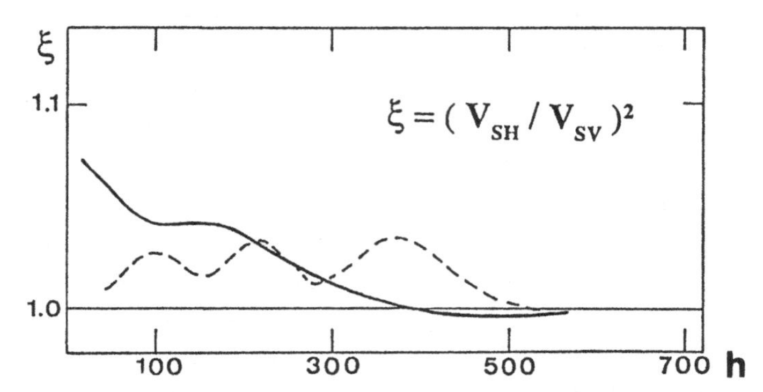

Figure 6-4 Square of the ratio of S_H and S_V velocities in North America and the Pacific versus depth h (km). Data are Love- and Rayleigh-wave phase velocities up to the second overtone for Love waves and the third overtone for Rayleigh waves (from Cara and Leveque, 1988). (AGU, copyright © 1988).

---

times of compressional teleseismic body waves, it is a common practice to estimate the lithosphere thickness variations from P residuals observed in different stations by assuming an *ad hoc* contrast of P velocity between the lithosphere and the asthenosphere. If anisotropy is present, it is clear that the thicknesses estimated from vertically incident rays depend on the orientation of the symmetry axes of the anisotropic materials in both the asthenosphere and the lithosphere.

To come back to surface-wave estimates of the lithosphere thickness, it is of interest to us to look at the S_V-velocity model of the Pacific obtained by anisotropic inversion of the Pacific data of Cara and Leveque (1988). Figure 6-5 shows both the isotropic shear-wave velocity model P7 fitting the Rayleigh-wave data alone (Cara, 1979) and the inverted anisotropic model for S_V-velocity fitting both the Love- and Rayleigh-wave data in the Pacific. Very small differences are found between the two models. The fact that anisotropy is taken into account in the inversion process has not produced any drastic change in the depth variation of the shear-wave velocity. This means that a little trade-off exists between shear-wave velocities and the anisotropic parameters in this multimode inverse problem where overtones up to the third were used both for Love and Rayleigh waves. More sophisticated models of the seismic anisotropy involving the azimuthal variation of velocities would be necessary to conclude on this point definitively but, at least with such a type of parameterization, it seems that the trade-off is not so severe, contrary to what was stressed by Regan and Anderson (1984).

The situation was quite different for the data used by these authors since only fundamental-mode data with a lower degree of resolution were considered. In their case it is clear that there is a much larger degree of freedom in the inverted parameters and that a strong trade-off may appear between poorly resolved parameters, yielding a possible trade-off between the lithospheric thickness and anisotropy. Nishimura and Forsyth (1989) have furthermore shown that the trade-off observed by Regan and

Anderson (1984) was directly linked to the way these authors parameterized the upper-mantle elastic model. Instead of using smoothly-varying velocity-depth models to start the inversion process, Regan and Anderson (1984) used a discontinuous representation of upper mantle velocity and anisotropy. This introduces strong non-physical *a priori* constraints in the depth variations of the model parameters, reducing artificially the number of degrees of freedom in the problem. Indeed, Nishimura and Forsyth (1989) who followed another philosophy while inverting similar data did not find any significant trade-off between the thickness of the lithosphere and anisotropy. This discussion on the trade-off between lithospheric anisotropy and lithospheric thickness is a clear illustration of what may happen when we interpret data with a poor intrinsic resolution in the framework of models presenting a very large degree of freedom. Reducing artificially the number of model parameters by an *ad hoc* parameterization may be physically misleading. The influence of anisotropy may then be important and change the conclusions related to the other parameters.

The use of multimode data which present a much better depth resolution enables us to look at the question of depth variation of the shear-wave velocity with much greater confidence. The robustness of shear-wave velocity in the models displayed in Fig. 6-5 strongly suggests that the seismic thicknesses of the oceanic lithosphere which were derived previously from fundamental Rayleigh waves in the framework of isotropic

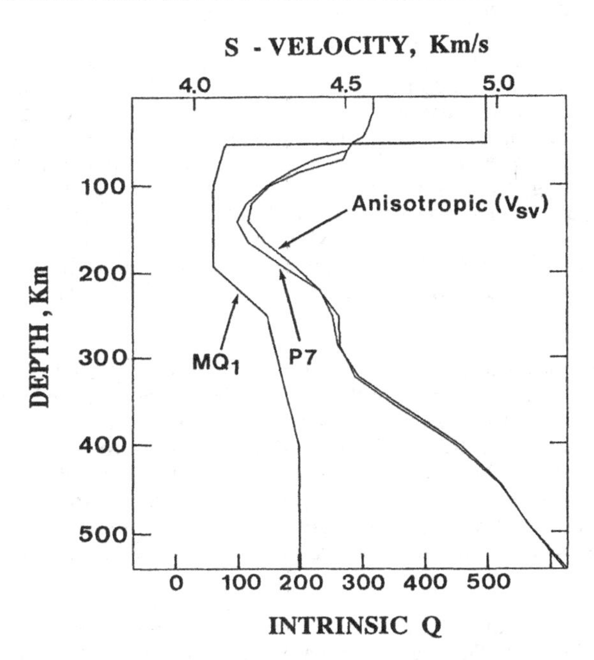

Figure 6-5 Anisotropic (S_V) and isotropic (P7) shear-wave velocity model of the Pacific (after Cara and Leveque, 1988). MQ1 is an attenuation model fitting the higher Rayleigh mode data (Cara, 1981). (AGU, copyright © 1988).

models should not be strongly affected when anisotropy is taken into account. As a direct consequence, this result means that the discrepancy between the thicknesses of the flexural and seismic lithospheres should not be significantly reduced when anisotropy is taken into account. If this result were confirmed by more sophisticated anisotropic modeling, it would mean that a time-dependent rheology, as already mentioned (Minster and Anderson, 1980), would be the more plausible explanation of the differences between the flexural and seismic lithospheric thicknesses.

These inferences on the depth extent of anisotropy are based on a few surface-wave dispersion data involving higher mode observations. These data are related to several thousand kilometer-long paths in the Pacific and North America (see Fig. 4-13 of Chapter 4). As they are related to waves propagating at close azimuths, they do not see possible azimuthal variations of velocities. The model of transversely isotropic structures we have used to draw the above conclusion is only valid for interpreting the data related either to intrinsically transversely isotropic structures or to long enough paths across structures representing different orientations of anisotropic material so that they can be considered as randomly oriented around the vertical direction. One possibility of the former model is provided by thin horizontal layers of partial melts embedded in an otherwise isotropic solid material as advocated initially by Schlue and Knopoff (1977). The latter model could correspond to a preferred orientation of the olivine crystal a-axes in the horizontal plane, but with random azimuths over the several thousand kilometer long surface-wave paths.

If a coherent flow direction exists in the mantle on such a large scale, the model used to make the above inferences may not be valid. As this may occur in the Pacific over the 10 000 km long paths displayed in Fig. 4-13 between the Tonga trench and the Vanuatu epicentral area and the western coast of the United States, Cara and Leveque have envisaged that a preferred orientation of olivine crystals might be coherent over the whole path which belongs to the same tectonic plate: the Pacific plate. This model is in agreement with the earliest finding of Forsyth (1975) who had established an azimuthal variation of Rayleigh-wave velocity with maximum velocity in a direction parallel to the absolute motion of the plate and was confirmed later by several authors, e.g. Tanimoto and Anderson (1985) from a study on the global scale, and Montagner (1985) in the Pacific.

Under the assumption that the cause of anisotropy is the preferred orientation of olivine a-axes parallel to the direction of the present-day motion of the Pacific plate, the two other axes being randomly oriented in the perpendicular plane, Cara and Leveque (1988) inverted the Love- and Rayleigh-dispersion curves, up to the third overtone by considering the percentage of oriented crystals as the unknown of the problem. No preferred orientation was assumed in this model for the other constituents of the upper mantle. The result is shown in Fig. 6-6 together with the inverted-model error bars (standard deviations of the model parameters in a Gaussian statistics for both the data and the model parameters).

It appears that olivine crystals have to be preferentially oriented down to nearly a 300-km depth to fit the data. This means that anisotropy is required by the model down to this depth, a conclusion which is similar to that reached in the framework of transversely isotropic models with a vertical axis of symmetry. 20% of olivine crystals in a pure-olivine mantle or 33% in a mantle made of 60% of olivine have to be preferentially oriented parallel to the present-day direction of the plate motion down to a

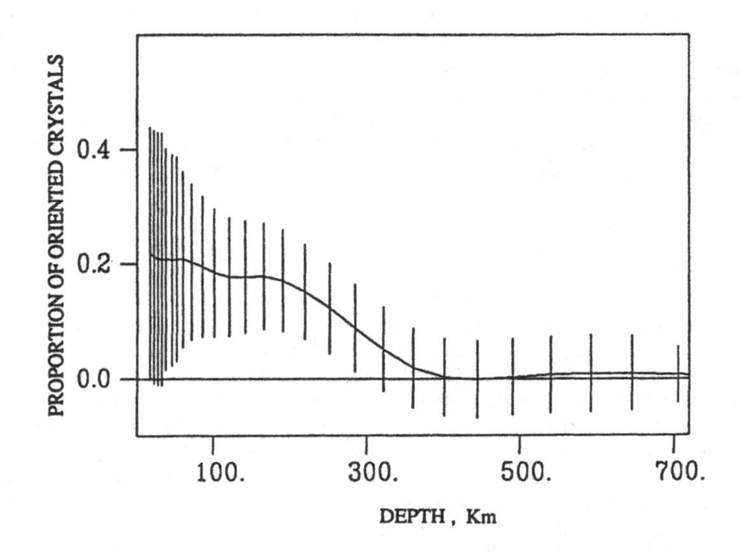

Figure 6-6 Depth variation of the proportion of olivine crystals oriented with their a-axes parallel to the absolute motion of the Pacific plate which is obtained from the Love- and Rayleigh-wave dispersion curves, including the higher modes (Cara and Leveque, 1988). Vertical bars are the standard deviations. Note that the data do not require anisotropy below a 300-km depth. (AGU copyright ©, 1988).

200-km depth to explain the observed Love/Rayleigh-wave discrepancy in this model. A more realistic model should include preferred orientation of the orthopyroxene content. The model displayed in Fig. 6-6 predicts azimuthal variations of surface-wave velocities over which we have no control since they sample a too narrow azimuthal range. To get information on azimuthal variations there is presently no possibility to work with overtones data. Further work based on the waveform inversion of individual seismograms (Cara and Leveque, 1987; Nolet, 1990) is one way to make some progress in this matter. We will describe farther the results obtained from the inversion of azimuthal variations of fundamental mode phase- and group-velocities, but before addressing the question of azimuthal anisotropy, let us review several more robust results obtained in the framework of transversely isotropic models.

6.1.2 Transversely isotropic models

Even if azimuthal variations of surface-wave velocity exist locally, the concept of a transversely isotropic medium is applicable to investigating the nature of anisotropy by averaging out the azimuthal variations. The perturbation theory applied to weakly anisotropic media by Smith and Dalhen (1973) and developed by Montagner and Nataf

(1986) provides the adequate theoretical framework to follow such an approach (see Chapter 2). Transversely isotropic models of the upper mantle can thus be used to describe equivalent transversely isotropic structures in regions broad enough to assume that the azimuthal anisotropy is averaged out, either due to some randomness in the physical symmetry of the medium or due to the fact that observations of seismic waves propagating in various directions are averaged out. Let us notice furthermore that transversely isotropic models are particularly well suited for describing the overall properties of global spherically symmetric Earth models since exact computations with decoupling between spheroidal and toroidal modes, as in isotropic models, are possible in this case.

We have already seen in Chapter 4 that global Earth model PREM (Dziewonski and Anderson, 1981) presents a weak anisotropy down to the 220-km Lehman discontinuity. Due to the rather poor depth resolution of the data used by Dziewonski and Anderson (1981) to build up this model, it was generally thought that the evidence of asthenospheric anisotropy was very poor on the global scale and that anisotropy confined to the lithosphere would explain the global data as well. Forming a second step in this approach are the global anisotropic models of Nataf et al. (1986) which are transversely isotropic with a vertical symmetry axis too but are laterally heterogeneous. In these models, a lateral heterogeneity was introduced by using two regionalization schemes, one based on a spherical harmonic expansion to angular order 6 and based on *a priori* fixed regions defined according to the age of the crust and its tectonic characteristics. Let us summarize the nature of anisotropy in the model based on the second regionalization scheme. The seven tectonic provinces defined by Okal (1977), four in the oceans and three in continents, were used in this approach. The best resolution of the ratio between S_H and S_V velocity was obtained in the depth range of 200 - 400 km. Nataf et al. (1986) found that S_H waves appeared to be faster than S_V waves in this depth range for the two regions of oceanic basins (ages between 30 and 135 million years). Young oceans (less than 30 million years) and old oceans (more than 135 million years) exhibit a different type of seismic anisotropy with S_V waves faster than S_H waves. A complicated pattern appears in the tectonic regions but, again, with S_H waves faster than S_V waves below a 200-km depth. In this study, no anisotropy was resolved beneath the shield and mountainous provinces at depths larger than 200 km.

At this stage, interpretation of the observed anisotropy strongly favors the preferred orientation of olivine crystals in the asthenospheric flow pattern. Horizontal orientation of olivine crystal a-axes under large horizontal shearing strain beneath oceanic plates should create azimuthal variations of surface-wave velocities similar to those displayed in Fig. 2-10. When averaged out over all azimuths, the equivalent transversely isotropic medium presents apparently too fast Love waves and S_H velocity larger than S_V velocity. An $S_H>S_V$ signature could then be characteristic of a predominantly horizontal flow inducing horizontal shearing of olivine-rich materials in the asthenospheric layer. On the other hand, regions of $S_H<S_V$ velocities should be typical of vertical shearing deformations induced by a predominantly vertical flow of upper mantle materials. This is what is observed in regions of oceanic ridges and in the old oceans where subduction and, possibly detached cold blobs of lithospheric materials could sink into the mantle.

In the Pacific ocean a strong anisotropy has been found in the asthenosphere by several investigators. It is interesting to present the recent study of Nishimura and

Forsyth (1989), who interpreted Love and Rayleigh waves criss-crossing the whole Pacific ocean in terms of both azimuthal anisotropy and transversely isotropic models. These authors applied the Tarantola and Valette inversion scheme together with the approaches of Smith and Dalhen (1973) and Montagner and Nataf (1986) to separate the transversely isotropic part of Love- and Rayleigh-wave velocity from their azimuthal variations. Rayleigh-wave phase velocities in the period range 20-120 s and Love wave phase velocities in the period range 30-120 s were used by these authors to derive a laterally heterogeneous fully anisotropic model of the Pacific. Five regions were defined according to the age of the crust to regionalize the dispersion curves. Assuming that a good azimuthal coverage was achieved in each of these regions they have inverted the resulting locally averaged dispersion curves in terms of transversely isotropic models. Contrary to the inversion performed by Regan and Anderson (1984), Kawasaki (1985) and, on the global scale, by Nataf et al. (1986), Nishimura and Forsyth (1989) have used a smoothly depth-varying S-velocity model as a starting model to avoid the introduction of artifacts in the depth-variations of the parameters. Both S_V velocity and the ratio of S_H to S_V velocities are relatively well resolved in this study to a 200-km depth (see Fig. 6-7).

The inverted anisotropic parameter ξ together with the S_V velocity are displayed in Fig. 6-8 and Fig. 6-9. Anisotropy extends well beneath the lithosphere and all the regions show a positive ξ value, that is a $S_H > S_V$ signature, compatible either with a fine

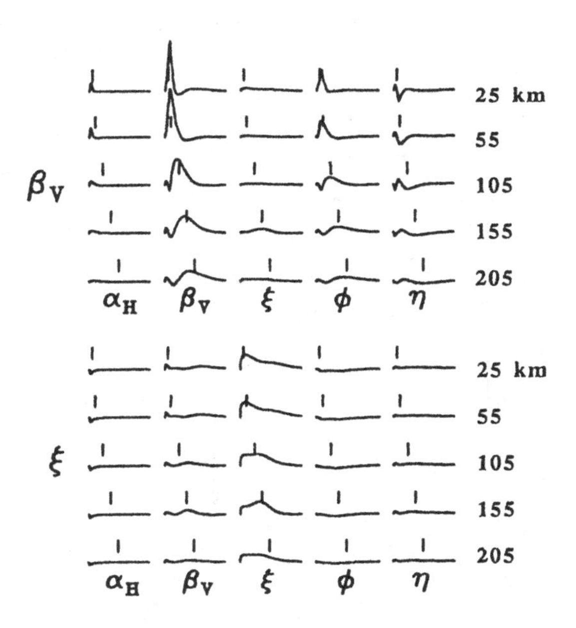

Figure 6-7 Resolution kernels for S_V velocity (β_V) and the square of the S_H- to S_V -velocity ratio (ξ). Tick marks show the depths where the resolution kernels are calculated (after Nishimura and Forsyth, 1989). (With permission of the Royal Astronomical Society, copyright © 1989).

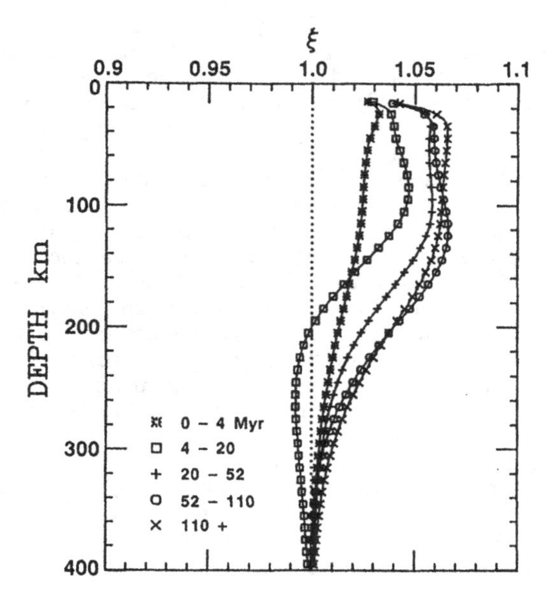

Figure 6-8 Depth variations of ξ in the Pacific (Nishimura and Forsyth, 1989). (With permission of the Royal Astronomical Society, copyright © 1989).

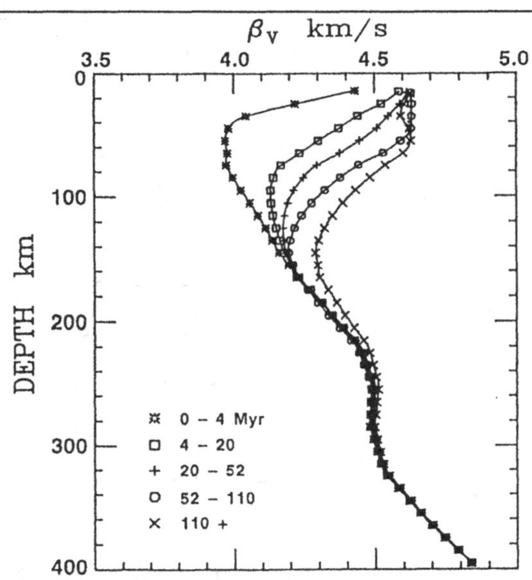

Figure 6-9 Shear-wave velocity models of the Pacific computed in the framework of an anisotropic model (after Nishimura and Forsyth, 1989). (With permission of the Royal Astronomical Society, copyright © 1989).

layering in the upper mantle with partial melting or horizontal flow of olivine-rich upper mantle materials. Note that the ξ values obtained by Nishimura and Forsyth (1989) in the depth range of 80-150 km are very close to those obtained in the Pacific from overtone data by Cara and Leveque (1989) (see Fig. 6-4), which correspond to the range of 52-110 million years.

6.1.3 Azimuthal anisotropy

Azimuthal variations of the Rayleigh-wave velocity were first observed by Forsyth (1975) in the Nazca plate. Since then, several investigators have tried to resolve a two-dimensional tomographic problem allowing for local azimuthal variations of phase- or group-velocity. Among them let us mention Kawasaki (1985) and Montagner (1989). These studies lead to the conclusion that Rayleigh waves display 1-2% of azimuthal variation in the velocity in the period range 50 to 200 s with fast direction parallel to the direction of the absolute plate motion. Love waves, on the other hand, do not provide clear azimuthal trends and the question is still open from the observational point of view.

On the global scale, Tanimoto and Anderson (1984) provided a first map of azimuthal variations of Rayleigh-wave velocities, showing a striking similarity between the direction of the return flow motion in the model of Hager and O'Connell (1979) of the mantle convection and the fast direction of Rayleigh waves. In this study Tanimoto and Anderson (1984) used a spherical harmonic approach. The map of azimuthal

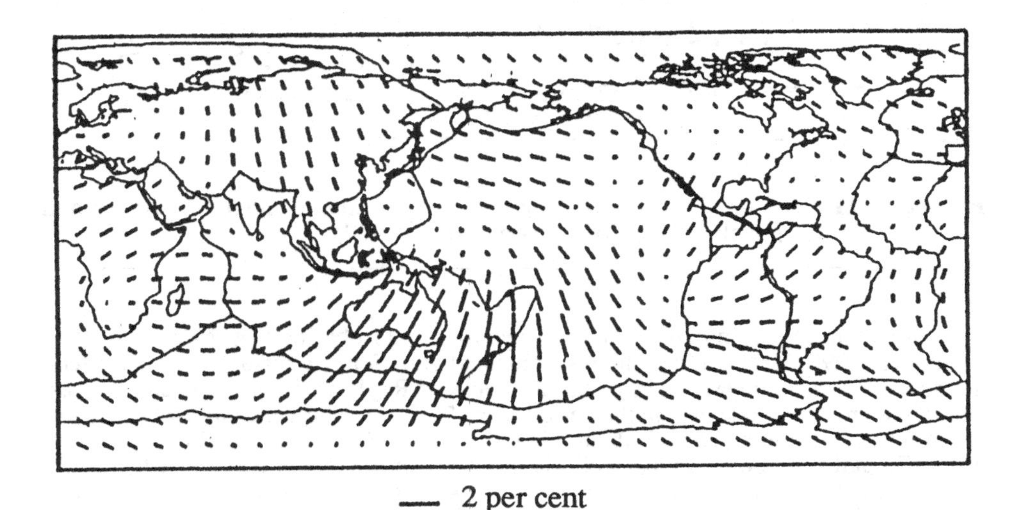

Figure 6-10 Azimuthal variations of a maximum Rayleigh-wave phase velocity at a period of 200-s obtained by Tanimoto and Anderson (1984) (spherical harmonic expansion to degree 3). (AGU, copyright © 1984).

variations of Rayleigh-wave phase velocity presented in Fig. 6-10 has been obtained to degree 3 and taking into account the angular dependence of phase velocity to degree 2 in the azimuth. Maximum azimuthal variations are less than 2%, attaining their maximum value in Australia, the south-east Indian Ocean and the south-east Pacific.

The map presented in Fig. 6-10 shows a striking resemblance with the flow lines obtained at a 260-km depth in the kinematic flow model of Hager and O'Connell (1979) shown in Fig. 6-11. In a model of the preferred orientation of olivine crystals in an overall horizontal shear strain, Rayleigh waves are fast in a direction parallel to the olivine a-axes. Comparison of Fig. 6-10 with Fig. 6-11 thus strongly suggests that the observed azimuthal variation of Rayleigh-wave velocity is due to the preferred orientation of olivine crystals in connection with the return flow pattern of plate tectonics in the upper mantle.

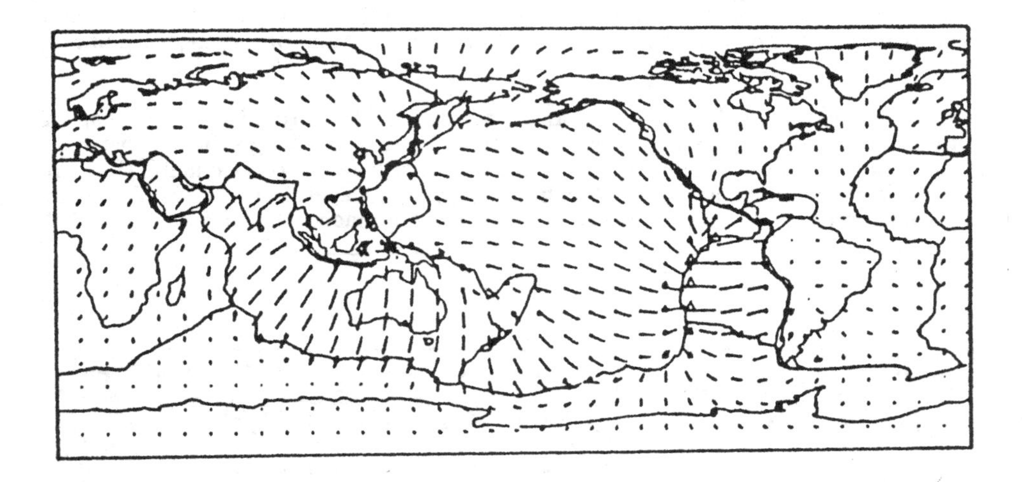

Figure 6-11 Flow lines at 260-km depth in the kinematic model of Hager and O'Connell (1979). (AGU, copyright © 1979).

Let us now summarize additional recent results obtained by Montagner and Tanimoto (1990, 1991) on the three-dimensional variations of anisotropic parameters. These authors have inverted a large set of Love- and Rayleigh-wave dispersion curves, using fundamental mode data in the period range 70-250 s. These data allow the authors to get information to a depth of about 450 km. A set of 2600 paths for Rayleigh waves and 2170 paths for Love waves, well distributed over the whole surface of the Earth, was analysed in order to retrieve the local dispersion data including their azimuthal variations. Instead of performing a direct three-dimensional inversion of the waveform data, as was done for example by Woodhouse and Dziewonski (1984), Montagner and Tanimoto have preferred to follow a two-step procedure. First, they have inverted the dispersion data related to the direct and inverse paths along great circles between a set of

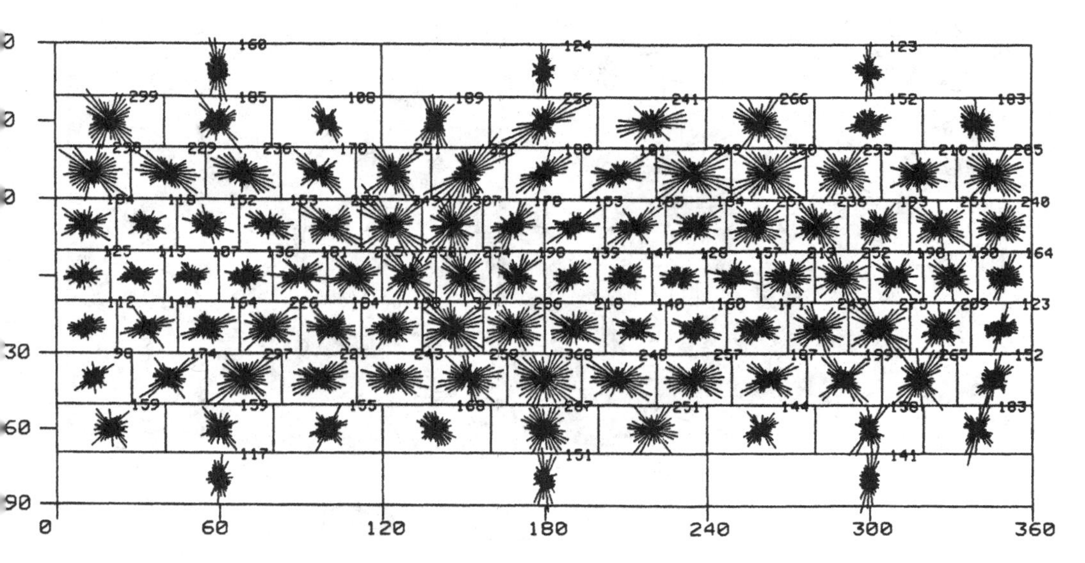

Figure 6-12 Azimuthal coverage of Rayleigh-wave rays in each 20° × 20° block for the data set used by Montagner and Tanimoto (1990) with the numbers of rays crossing the block. (AGU, copyright © 1990).

epicenters and seismic stations in terms of regional, azimuthally varying velocities using an iterative simplified two-dimensional inversion scheme. Second, they have performed an inversion of the local dispersion data along the vertical coordinate using the least-square procedure of Tarantola and Valette (1982).

Due to the large number of epicenter-station and great-circle path dispersion curves used by Montagner and Tanimoto (1990), a very good coverage of the Earth's surface is obtained. Dividing the surface of the Earth into 20° × 20° blocks, the hit count of rays per block is between 98 and 350 for Rayleigh waves with a good azimuthal distribution (Fig. 6-12). A similar situation is obtained for Love waves with a hit count per block between 87 and 339. This means that both the local average velocities and their azimuthal variations can be obtained unambiguously if the *a priori* correlation length is set to a large enough value (see sections 4.1 and 4.2). Unfortunately, since the amount of great circle data to be inverted was too large, Montagner and Tanimoto (1990) did not use the well controlled and consistent algorithm of Tarantola and Valette (1982) for inverting the observed dispersion curves. They developed an approximate iterative method, similar to the Simultaneous Iterative Reconstruction Algorithm (Humphreys and Clayton, 1988) which converges to the least square solution. Doing so, Montagner and Tanimoto (1990) first obtained the different terms of the local azimuthal expansion of phase velocities for both the Love and Rayleigh waves (i.e. the 0φ, 2φ and 4φ terms). Outputs of this first inversion step can be presented in global maps of

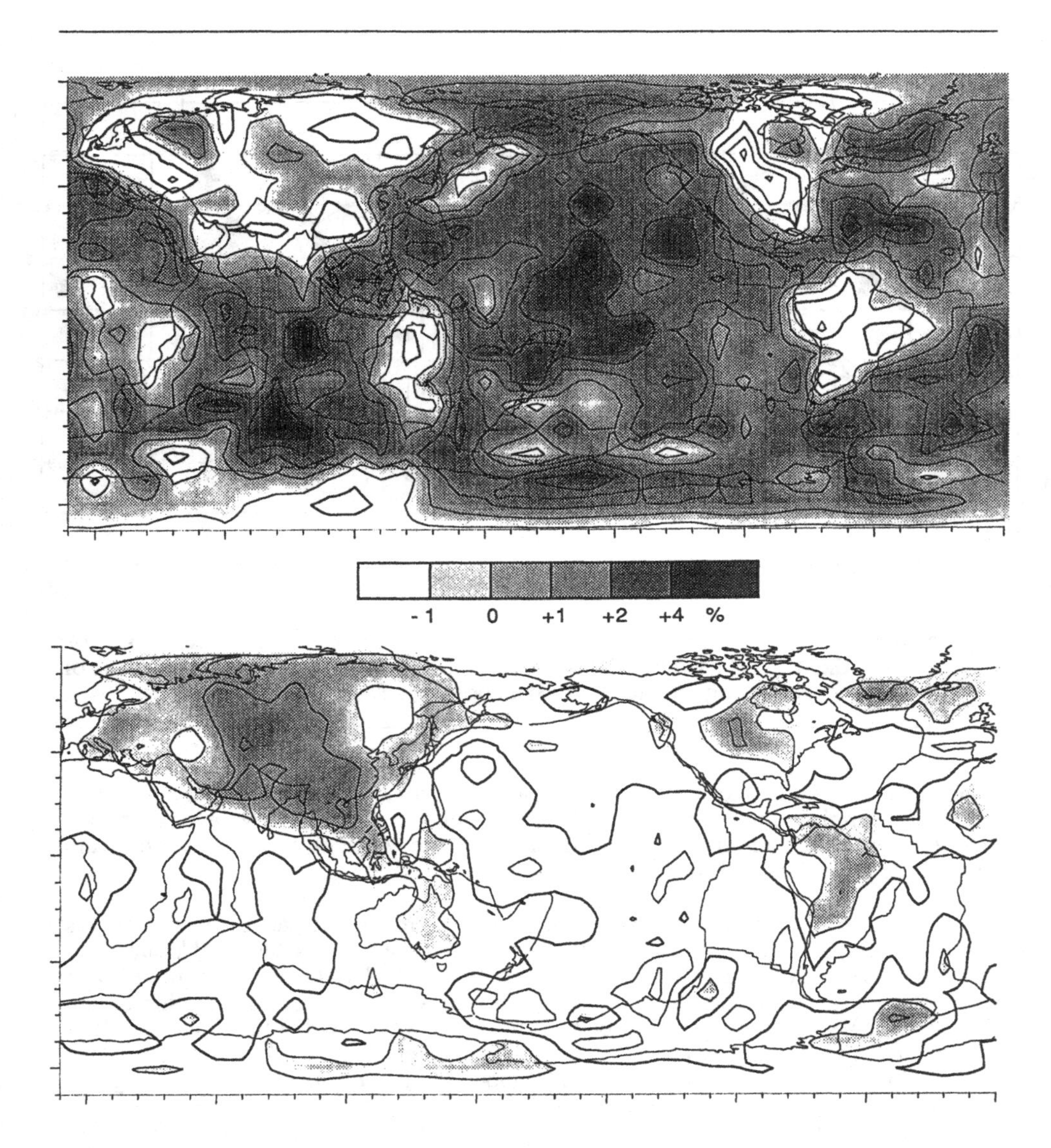

Figure 6-13 Relative variations of the transversely isotropic term ξ at depths of 100 km (a) and 310 km (b), obtained by Montagner and Tanimoto (1991). Positive ξ values beneath large oceanic regions indicate S_H velocities faster than S_V velocities and are indicative of a predominantly horizontal flow of an olivine-rich material. A similar pattern is found at greater depths beneath large continental regions.

anisotropic phase velocities for Love and Rayleigh waves at different periods.

Different tests with synthetics were performed by the authors to test the reliability of the inverted phase velocity variations over the surface of the Earth and, in particular, that of the azimuthal velocity variations. As a general conclusion, it appears that the azimuthally isotropic terms are more robust than the higher order terms. One of these tests was designed to check whether a realistic variation of an isotropic phase velocity over the surface of the Earth could induce artificial azimuthal velocity variations. Average theoretical phase velocities along great circle paths were first calculated for the isotropic distribution of velocity and were then inverted in terms of anisotropic velocities. Ideally no azimuthal variations of velocities should show up at the end of this experiment. Actually some azimuthal variations are found indicating that there is a trade-off in these maps between lateral heterogeneity of the phase velocity and its azimuthal - or anisotropic - behavior. Those artificially induced azimuthal variations, however, are smaller by a factor of two than the azimuthal variation of velocity obtained from the inversion of real data. Moreover, the directions of maximum velocity obtained in this experiment are very different from those obtained with real data (Montagner and Tanimoto, 1990) which give us some confidence in the output of the inversion. Nevertheless this experiment shows us that azimuthal variations of phase velocity are much less constrained than the transversely isotropic terms.

Actually, the observed azimuthal variations of Rayleigh-wave velocity are found to be quite constant between periods of 70 and 130 s where they attain about 1.3%. At longer periods they decrease and amount to only 0.3% at 250 s, being thus smaller than in the previous finding of Tanimoto and Anderson (1984). The direction of fast Rayleigh waves coincides well with the direction of the absolute motion of the tectonic plates providing further support to the idea that the preferred orientation of olivine crystals might be the main cause of the observed anisotropy.

Once the phase velocity maps were obtained, a second inversion was performed along the vertical coordinate in order to get the three-dimensional image of the upper mantle. Montagner and Anderson (1989a, 1989b) showed that the Rayleigh 4φ and Love 2φ terms were small in view of several possible compositions of the upper mantle assemblages. For this reason, Montagner and Tanimoto (1991) only considered the Rayleigh 2φ and the Love 4φ terms. The main results they have obtained can be summarized as follows:

• Lateral heterogeneities of shear-wave velocities are confined to the upper 250 - 300 km depths.

• Departure from isotropy is resolved down to a depth of 200 to 300 km. There is no direct evidence of large-scale anisotropy below this depth.

• S_H waves are faster than S_V waves at a depth around 100 km beneath large oceanic plates such as the Pacific. A similar pattern although with smaller amplitude is found at a greater depth beneath large continental regions (Fig. 6-13).

• Directions of fast S waves appear to correlate well with the direction of the absolute motion of the plate, the best correlation being achieved at a 200-km depth.

The global study of Montagner and Tanimoto (1990, 1991) gives us a clear confirmation of the fact that the upper mantle beneath the Pacific is very likely to be anisotropic to a depth of at least 200 km. This depth largely exceeds the lithosphere thickness of the Pacific and even if the large anisotropy of the lithosphere contributes to this anisotropic signature, these results clearly confirm that the asthenosphere is anisotropic.

Due to the apparent simplicity of upper mantle anisotropy in the Pacific, let us come back to the surface-wave study of Nishimura and Forsyth (1989), by looking now at the inferences they made on the origin of the azimuthally varying terms. These authors followed the approach of Montagner and Nataf (1988) and inferred the depth variation of the 2φ azimuthal terms Gc, Bc and Hc (see section 2.5). Since parameter $G_c=(C_{55}-C_{44})$ presents partial derivatives which are identical to those of β_V for Rayleigh waves, it is the best resolved parameter among all the azimuthal terms.

Figure 6-14 shows the 2φ anisotropic term G_c versus depth for two regions dealt with by Nishimura and Forsyth (1989). Figure 6-14-a shows the depth variation of this parameter for regions younger than 80 million years. The azimuthal anisotropy extends down to approximately 200 km, with a maximum around 75 km. A simple uniform

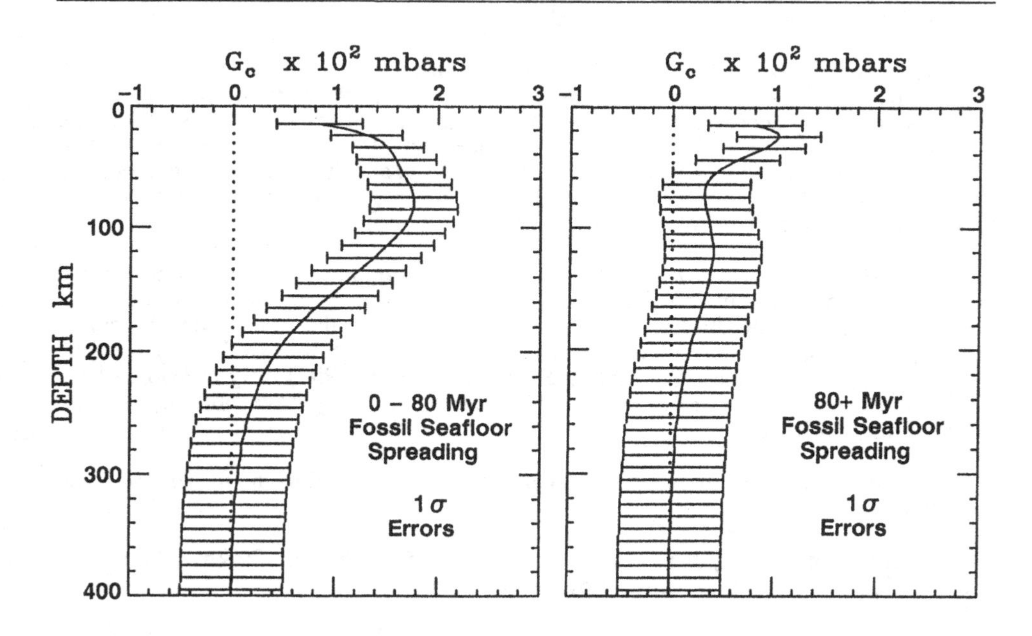

Figure 6-14 Azimuthal anisotropy parameter G_c versus depth for the region of the Pacific younger than 80 million years (left) and older than 80 million years (right) (after Nishimura and Forsyth, 1989). (With permission of the Royal Astronomical Society, copyright © 1989).

pattern of the azimuthal anisotropy is thus likely to be the dominant anisotropic feature of the lithosphere-asthenosphere system in the young regions of the Pacific. On the other hand, in the oldest regions of the Pacific, the parameter G_c appears to be zero in the asthenosphere and the lower lithosphere (i.e. at a depth larger than 50 km), while it differs from zero in the upper lithosphere. Nishimura and Forsyth have noticed that the direction of the present day motion of the plates coincides with its fossil direction for ages less than 80 million years while it differs for older ages. Starting from this fact, they give a simple explanation of the differences in depth variations of G_c for the two different provinces. Due to a lack of resolution, no clear azimuthal pattern is found in the oldest part of the Pacific since the direction of the asthenospheric flow has changed with time. On the contrary, in the young Pacific the present-day motion is parallel to the direction of the motion at the time the lithosphere was formed. A uniform anisotropic pattern could thus develop in the province aging less than 80 million years. This interpretation of the depth-distribution of Gc in the oldest part of the Pacific is in full agreement with the plate thickening model discussed in Chapter 5.

6.1.4 Possible causes of seismic anisotropy in the asthenosphere

There are two main models of the upper mantle composition. The classical pyrolite model by Ringwood (1975) is made of 59% olivine, 17% orthopyroxene and 12% garnet, e.g., Estey and Douglas (1986). This model fits the existing seismic velocity distribution of the upper mantle according to Weidner (1985). An alternative model proposed by Bass and Anderson (1984), named the "piclogite" model, is made of 18% olivine, 6% orthopyroxene, 37% clinopyroxene and 39% garnet. In an attempt to discriminate between the two models on the basis of average anisotropic properties of the upper mantle, Montagner and Anderson (1989a), computed a large family of macroscopic elastic parameters for both compositions. Their main conclusion is that the pyrolite model is clearly favored in the upper part of the mantle above a 220-km depth while the presently available seismological data are insufficient to discriminate between the two models in the depth range 220-400 km. Among the anisotropic constituents of pyrolite, olivine is clearly the major component and should thus play the main role in the anisotropic properties of the asthenosphere.

At high temperatures, Fig. 5-6 by Nicolas and Christensen shows that axis [100] of olivine is oriented in the main shearing direction being parallel to the flow lines while axes [010] and [001] form a girdle in the perpendicular plane. Such temperature conditions are typically asthenospheric conditions so that an hexagonal symmetry with the symmetry axis oriented parallel to the flow direction of asthenospheric material might be the main cause of observed seismic anisotropy in the depth range 100-200 km. Beneath large oceanic plates the overall flow direction should be mainly horizontal thus yielding horizontal axes of symmetry. When averaged out in different azimuths, this model is fully compatible with surface wave observations which show that S_H waves are faster than S_V waves in the the depth range 100-200 km beneath oceanic plates. This model also predicts azimuthal variations of both Rayleigh- and Love-wave velocities. Although the former azimuthal variations of velocities are clearly observed, this is not yet the case for Love waves so that additional investigations are necessary. Either the other constituents of the upper mantle, such as orthopyroxene which has a diluting effect on anisotropy, play a role or other sources of anisotropy have to be taken into account.

The main alternative to the model of preferred orientation of olivine crystals is provided by shape-induced anisotropy due to the geometry of partial melts in the asthenosphere. If partial melts were predominantly organized in flat horizontal films, the overall anisotropy would not exhibit any azimuthal variations in the horizontal plane but could well explain the fact that S_H waves are faster than S_V waves.

Beneath continents the situation is much less clear. In Europe, Wielandt et al. (1987) and Maupin and Cara (1991) also found evidence for high S_H velocity around 100-km depth. In several provinces of Africa, Hadiouche et al. (1989) found evidence for a high S_H/S_V velocity ratio at a depth greater than 50 km. These observations could be accounted for by horizontal shearing deformation of olivine-rich material at the base of the lithosphere. At greater depths, Leveque and Cara (1985) obtained from surface-wave overtones a positive S_H/S_V-velocity ratio beneath North America to a 400-km depth. The recent global inversion of fundamental mode surface-wave data by Montagner and Tanimoto (1991) also shows that this could the case beneath most continental areas at a 370-km depth.

Although it appears to be weaker than beneath oceans, the seismic anisotropy of the asthenosphere beneath the continent might share the same origin: i.e. the preferred orientation of olivine [100] axes in the horizontal plane under the action of the present shearing deformation of asthenospheric materials. What is sometimes referred to as the continental drag might thus induce an observable seismic anisotropy at a greater depth than beneath the ocean due to the larger lithospheric thickness of continents.

6.2 Seismic anisotropy at greater depths

At depths greater than 400 km both seismological investigations and speculations based on mineralogical data show that there are several regions of the Earth where seismic anisotropy may exist but it is clear that more investigations are necessary to confirm what presently still relies mainly on speculations. The first problem is that deep anisotropy causes only small anomalies in the seismic data. Large heterogeneities of the upper mantle and also anisotropic effects of the lithosphere and asthenosphere have first to be corrected for before the observations can be interpreted in terms of deep anisotropy. If we want to use some constraints based on mineralogical data, the other problem is that the elastic properties of high-pressure and high-temperature phases are poorly known as far as anisotropy is concerned. A further difficulty lies in the uncertainties we have about the mineralogical composition of the transition region and the lower mantle. In the Earth's core, the situation is still more problematic since very little is known about anisotropy of the high-pressure iron phases.

6.2.1 Upper-mantle transition region and lower mantle

The upper-mantle transition region corresponds to the zone of a large seismic -velocity gradient between the bottom of the low-velocity layer and the top of the lower mantle, roughly between 300 and 700 km. In traditional seismic velocity models presenting two discontinuities at 400 and 650 km, the upper mantle transition region is generally restricted to the layers lying between the two discontinuities. This region of the Earth was initially introduced by Bullen who called it C region. This zone of a large velocity gradient is necessary to account for the decrease of the slope of the P-wave travel time curve around 20° epicentral distance (Bullen and Bolt, 1985). While the lower mantle corresponds to the seismic velocity distribution well explained by self-compression of a homogeneous material in adiabatic equilibrium, this is not the case for the upper-mantle transition region which was initially interpreted as a region of phase changes. Actually, seismologists and petrologists tried to constrain the upper mantle composition by comparing the observed seismic velocity distribution with curves predicted by different mineral assemblages (Anderson, 1989). It appears that while some consensus of opinions exists that olivine β-spinel phase change plays an important role in the strong velocity gradient observed around 400 km, the situation is much less clear at greater depth; and the possibility of a chemical change was advocated, for example, by Bass and Anderson (1984) or by Duffy and Anderson (1988).

The transition region appears in fact as a laterally heterogeneous zone where details of the seismic velocity distribution are poorly known. Even the number of seismic discontinuities to be considered is still the subject of debate. Both seismological progress and additional seismic laboratory experiments are needed to constrain better the physical and chemical state of this region of the Earth. Concerning seismic anisotropy, to date there is no direct evidence. All surface-wave models, either from higher mode data (Cara and Leveque, 1988; see Fig. 6-5 and Fig. 6-6) or from long period fundamental mode data (Montagner and Anderson, 1989b; Montagner and Tanimoto, 1991) show no resolvable anisotropy in the depth range 400-600 km.

Depending on the type of mineralogy in the upper mantle transition region, a possibility of anisotropy is either favored or not. It is clear that if the high pressure phase of olivine, β-spinel, possesses a strong anisotropy with up to 24% of shear-velocity variations, γ-spinel is much less anisotropic. In the cold subduction zones, the high pressure form of orthopyroxene, clinopyroxene and garnet may give a highly anisotropic material (Anderson, 1989).

In summary, due to the uncertainties still prevailing in the depth range of the transition region, it is difficult to make any conclusion. There is no direct evidence of large-scale seismic anisotropy there but, at least in subduction zones it is very likely that seismic anisotropy may exist on a smaller scale in connection with high local strain fields.

Below the transition zone there is no seismological evidence for any seismic anisotropy in the lower mantle, except perhaps in the lowermost 200 km of the mantle, near the core-mantle boundary (D" region). Actually, if significant seismic anisotropy would exist in the whole lower mantle, it is likely that it would produce large observable effects on mantle shear waves. From laboratory data, it is clear that if perovskite - the major constituent of the lower mantle - presents orthorhombic symmetry and a rather large elastic anisotropy under ambient conditions (e.g. Table 3-2), the situation may be very different under the lower mantle conditions. Wang et al. (1990), who analysed the microstructure of perovskite crystals synthesized at 26 GPa and 1600°C, found a cubic structure. Perovskite might thus exhibit a cubic symmetry and a small elastic anisotropy under the lower mantle conditions.

6.2.2 Core-mantle boundary

The core-mantle boundary at a 2900-km depth is probably the most chemically active transition zone within the Earth with high temperature fluid iron in contact with a high pressure phase of solid silicate, most probably perovskite (for a review, see Jeanloz, 1990). The lowermost 200 km of the mantle - or D" region according to Bullen's nomenclature is presently the target of intensive reasearches due to its great importance for the mantle and core dynamics. Region D" appears as a strongly heterogeneous region where a negative velocity gradient is expected (e.g. Young and Lay, 1990).

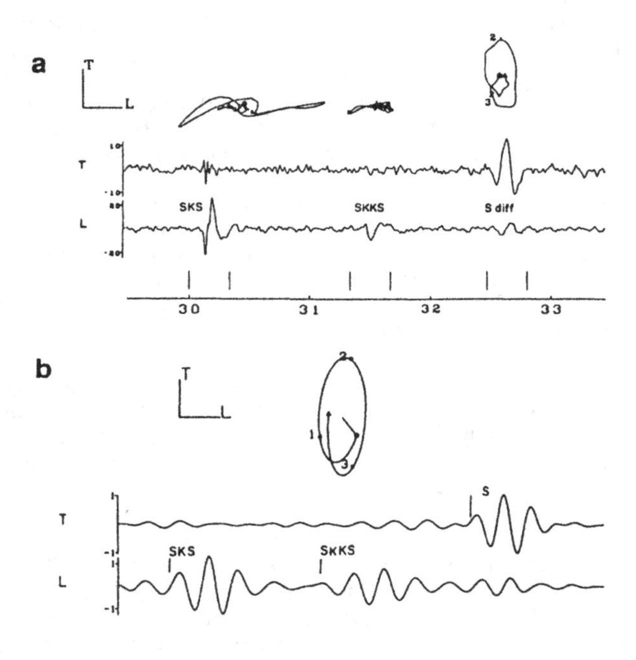

Figure 6-15 Radial (L) and transverse (T) records of a deep earthquake observed at the Geoscope station WFM, Massachusets, at 117° epicentral distance: a) raw broad-band record, b) data filtered in the bandwidth 0.055-0.77 Hz. The abnormally large amplitude of the transverse component of the diffracted S_V might indicate seismic anisotropy in region D" (after Vinnik et al., 1989b). (AGU, copyright © 1989).

The possibility of seismic anisotropy in region D" has been mentioned several times. Doornbos et al. (1985) and Cormier (1986) discussed the possibility of a transversely isotropic D" region (with vertical axis of symmetry). More recently, Vinnik et al. (1989b) have suggested that D" might be azimuthally anisotropic in order to explain abnormally too large diffracted S_V waves observed in the S_H shadow zone of the core at epicentral distances between 107° and 117° (e.g. Fig. 6-15). The possibility of a shear-wave splitting within the lithosphere or asthenosphere beneath the station can be ruled out because no significant splitting is observed on the SKS wave. As a negative velocity gradient can explain the large amplitude of the observed diffracted S_V around 110° as well (Kind and Muller, 1977), the question of anisotropy of the D" region remains largely open. Further observations and theoretical developments are necessary to model the diffracted S waves in the shadow zone of the core.

6.2.3 Inner core

Two independent sets of seismological observations have led seismologists to consider seriously a possibility that the inner core might bear a strong anisotropy with hexagonal symmetry and a symmetry axis coincident with the rotation axis of the Earth.

The first observations come from PKIKP arrival-time residuals. While analysing a large set of PKIKP travel-time residuals, Morelli et al. (1986) selected observations between 170° and 180° which correspond to rays with a nearly vertical incidence angle within the inner core. In this study, the effects of the upper-mantle travel-time anomalies, together with the effect of the core-mantle topography, were removed from the data analysed in terms of the spherical harmonic expansion. A strong zonal character appears in the residual inner-core travel times, attaining about 2 s: vertically propagating PKIKP waves are faster along the rotation axis of the Earth than in the equatorial plane. As no plausible heterogeneity can explain these observations, Morelli et al. (1986) proposed that the inner core might be anisotropic with a cylindrical symmetry parallel to the Earth's axis of rotation. An average 1%-velocity anisotropy of the inner core seems to be required by the data. Independent confirmation of this result was provided by Shearer et al. (1988) and Shearer and Toy (1991).

The second observation comes from the analysis of split core modes in the free oscillation spectra of the Earth (Woodhouse et al, 1986). These data need 25-km high undulations of the inner-core boundary and large heterogeneities near the upper part of the inner core to be explained by isotropic models (Giardini et al., 1987). A homogeneous spherical model presenting a hexagonal anisotropy of the inner core with the axis of symmetry coincident with the Earth's rotation axis is an alternative model which fits the splitting observations as well.

Interpretation of these observations has been made in terms of preferred orientation of a high pressure phase of iron crystals (hexagonal closely packed phase or ε-phase; see Anderson, 1986). These hcp crystals are believed to be anisotropic with hexagonal symmetry (Jephcoat et al., 1986), although their elastic properties have not been observed directly at the high pressure-temperature conditions of the inner core but have been extrapolated from a titanium analog. These crystals might have been preferentially oriented parallel to the Earth's rotation axis by crystal growth during the accretion process of the inner core. They could also reflect a present solid-state convection of the inner core (Jeanloz and Wenk, 1988).

From the seismological point of view, the question of the inner core anisotropy once again illustrates the difficulty of separating the effects of heterogeneity from anisotropy. A recent seismological investigation based on PKiKP-PcP observations (Souriau and Souriau, 1991) has proven that the shape of the inner core is closer to a perfect sphere than what was predicted by Giardini et al. (1987). Consequently, the observed travel-time anomalies of PKIKP cannot be explained by an abnormally large ellipticity of the inner core boundary. On the other hand, core-mode observations made from SRO stations (Suda and Fukao, 1990), recently confirmed by data of a superconductive gravimeter, do not support the splitting observations of Woodhouse et al. (1986) and Giardini et al. (1987). They are compatible with a spherically symmetric isotropic inner core.

As for other deep regions of the mantle, seismic anisotropy of the inner core can be considered as a still open question. It is clear that additional seismic observations of both body-wave travel times and free-oscillations spectra, together with better physical constraints on the high-pressure iron phases will be necessary before we can make any conclusive statement on the reality of the inner-core anisotropy.

Chapter 7 - Conclusions and outlook

It has been known for a long time that minerals and many sedimentary and crystalline rocks are elastically anisotropic. Due to geodynamic processes and/or action of the gravity field, many rock samples show strong fabrics. Crystallographic axes of minerals are for instance frequently aligned to particular petrofabric structures. Such structures can in turn produce a seismic anisotropy which is observable by direction-dependent velocities or by other seismic anomalies related to the wave polarization.

From the strict seismological point of view, the following observations strongly suggest the presence of seismic anisotropy in different parts and depths of the Earth (Table 7-1):

- Shear-wave splitting observed with a typical S-velocity anisotropy around 4-5%. The observed splitting is mainly interpreted as due to oriented systems of cracks in the upper crust and due to preferred orientation of olivine crystals in the subcrustal lithosphere where the main source of splitting of SKS and ScS waves is believed to originate.

- Azimuthal dependence of P velocities. In the subcrustal lithosphere beneath the oceans the P-velocity anisotropy varies between 3 and 8% and beneath the continents between 2 and 7%. Most of these observations have been interpreted as being due to preferred orientation of olivine crystals.

-Systematic spatial variations of teleseismic P residuals at seismological stations. An average P-velocity anisotropy around 8-9% in the continental subcrustal lithosphere, derived from crossing seismic waves, agrees with the anisotropy determined on specimens of ultramafic rocks (peridotites).

- S_H-S_V anisotropy as deduced from surface wave studies. Azimuthally averaged S_H velocity is almost always larger than S_V velocity for the deep oceanic lithosphere. Beneath the continents, the ratio of both velocities varies.

- Azimuthal variation of Rayleigh-wave velocities, up to about 2% in the oceans.

All the observations mentioned above have been considered as direct signs of seismic anisotropy in several regions of the Earth, where alternative explanations in terms of isotropic lateral variations of seismic velocities can be ruled out. In these regions, either the slope of seismic discontinuities are small over the investigated area, or lateral variations of seismic velocities are small enough, so that considerations of a trade-off between anisotropy and lateral heterogeneities do not provide a reasonable explanation for the observed seismic anomalies.

7.1 Possible causes of the observed seismic anisotropy

The crack-induced anisotropy has been observed in the upper parts of both the oceanic and continental crust. This type of anisotropy is widely modeled by systems of

Table 7-1 Present state of knowledge of seismic anisotropy

	Depth range	Seismic observations	Origin of anisotropy
CRUST			
Continents:			
Sedimentary Basins	~ 0 - 5 km	P_H/P_V-velocity differences	Layering of sediments
		S-wave splitting	Vertical cracks
Upper crystalline crust	~ 0 - 15 km	S-wave splitting Lg waves	Vertical cracks and fractures Foliation of rocks
		Reflections on faults	Anisotropy of mylonites
Lower crust	~ 15 - 30 km	Horizontal reflectors	Several possible models: - laminated structures - horizontal shearing - horizontal cracks with fluids
		No direct evidence	Coherent crystalline anisotropy over large volume
Oceans:	~ 5 - 11 km	Borehole data S-wave splitting Azimuthal variations of P velocities	Layering of sediments Vertical cracks and fractures
SUBCRUSTAL LITHOSPHERE			
Continent:	~ 30 - 150 km	Pn-azimuthal variations Long-range profiles P-wave residuals SKS and ScS splitting	Preferred orientation of olivine and orthopyroxene, either frozen-in or reoriented within a tectonic strain fabric
Ocean:	~ 10 - 100 km	Pn-azimuthal variations Long-range profiles P-wave residuals SKS and ScS splitting Love/Rayleigh-wave incompatibility	Frozen-in preferred orientation of olivine and orthopyroxene

ASTHENOSPHERE			
Continent:	~ 150 - 400 km	Love/Rayleigh-wave incompatibility	Orientation of olivine in the present-day flow?
Ocean:	~ 50 - 300 km	Love/Rayleigh-wave incompatibility Azimuthal variation of Rayleigh-wave velocity	Orientation of olivine in the present-day flow?
UPPER MANTLE TRANSITION REGION			
	~ 300 - 700 km	No clear evidence	Anisotropy in subducted slabs due to mineral orientation?
LOWER MANTLE			
	~ 700 - 2600 km	No evidence	
D" REGION			
	~ 2600 - 2900 km	Splitting of diffracted S waves?	Not known
INNER CORE			
	~ 5154 - 6371 km	Splitting of core modes? PKIKP-wave residuals?	Preferred orientation of high pressure iron phase?

parallel vertical microcracks or fractures which may be equivalent to a transversely isotropic medium with a horizontal symmetry axis. Such anisotropy in combination with a subhorizontal layering or a schistosity may yield an orthorhombic or lower-symmetry system. Only a few seismic experiments indicate the presence of seismic anisotropy in the lower crust, though the results of the reflection seismology suggest a large-scale layering (equivalent to a transverse isotropy with the vertical symmetry axis for large wavelengths). In the crust, mylonite zones represent regions of an extremely high seismic anisotropy (over 20% for P velocities). This can serve as a model for seismic reflectivity of faults within the crystalline crust.

Seismic anisotropy of the subcrustal oceanic lithosphere is very likely to be caused by a frozen-in alignment of olivine crystals. Crystallographic a-axes appear to be oriented horizontally, parallel to the direction of the upper mantle flow at the time the lithosphere was formed by cooling. Changes in directions of the fast Pn velocity which coincide with the olivine a-axis direction, and a weak azimuthal dependence of Rayleigh-wave velocity in the oldest part of the Pacific both suggest that seismic anisotropy has been affected by changes in the direction of the Pacific plate motion during the last 200 million years. Subcrustal lithospheric anisotropy might thus keep the memory of a past mantle flow.

The subcrustal oceanic lithosphere most probably retained its seismic anisotropy in the large volume of subducting slabs due to the short time constant involved in the subduction process. Furthermore, additional anisotropy might be formed by shearing strain near the contact of the subducting slab with the lithosphere above the subduction. This makes the presence of seismic anisotropy very likely in the deep continental lithosphere resulting from accreted active margins. Indeed, the anisotropy patterns in the subcrustal lithosphere beneath continents indicate more complicated olivine orientations. In constrast to the oceanic lithosphere, the velocity extremes seem to be inclined and the anisotropic structures may represent relics of paleosubductions of an ancient oceanic lithosphere. The anisotropy pattern of the continental lithosphere is even more complicated when it is modified by later tectonic processes like rifts and large-scale shearing zones. Continents exhibit complicated structures which contribute to the memory of the Earth.

Most of the observations of seismic anisotropy in the asthenosphere come from surface waves. In contrast to the frozen-in anisotropy of the lithosphere, the seismic anisotropy in the asthenosphere reflects a present-day convection flow producing a preferred orientation of crystals, mainly olivine. Azimuthal variation of very long period Rayleigh waves is a key observation supporting this model. Partial melt might also contribute to the asthenospheric anisotropy if melted material could be preferentially oriented in thin horizontal layers. Presently available data suggest that large-scale seismic anisotropy in the mantle beneath a depth of about 300 km is very weak.

Several studies of travel times and free oscillations suggest that seismic anisotropy might exist in the inner core. Anisotropy representing a hexagonal symmetry, with the main axis oriented along the axis of the Earth's rotation, has been tentatively interpreted as being caused by a strong alignment of hexagonal close-packed iron. More investigations, based both on seismic observations and laboratory experiments are needed to solve the problem.

7.2 Outlook

Seismic anisotropy is becoming a more and more important subject in various fields of seismology and seismic prospecting due to its ability to map directional features of the Earth's structures. Since 1985, an increasing number of exploration seismologists and reservoir and production engineers have been aware of the potential importance of shear-wave splitting in the crust, due to its ability to give some information on the anisotropic properties of physical parameters which could be linked with permeability. Earthquake seismologists are now beginning to investigate the anisotropic properties of the upper crust as an indicator of the present state of tectonic stress and it is not impossible that, in the future, well monitored observations of changes in the anisotropic properties of the seismogenic layers of the crust could be used as an earthquake precursor.

Concerning the greater depths of the Earth on which this monograph is mainly focused, seismic anisotropy can yield very valuable information on the orientation of crystals and structures. Seismic anisotropy can become a perspective method for studying the tectonics of the Earth's interior (Fig. 7-1). But this requires closer relations between structural geologists and geophysicists, as well as a better understanding of the relationships between large-scale fabric and its representation in terms of seismic anisotropy. In order to retrieve information on anisotropy, including orientations of the symmetry axes, future seismic experiments with closely spaced and mostly three-component stations have to be used, as both P and S velocities and shear-wave polarizations in many directions are required. The theory of wave propagation in anisotropic media is well developed and technical tools for extracting adequate information are available. Interpretation of seismic observations in terms of laterally heterogeneous and anisotropic structures is one of the main challenges of structural seismology in the future. High quality observations from well designed seismic experiments together with advanced theoretical methods - including well controlled inverse methods and waveform modeling techniques - will be necessary to make decisive steps in this direction.

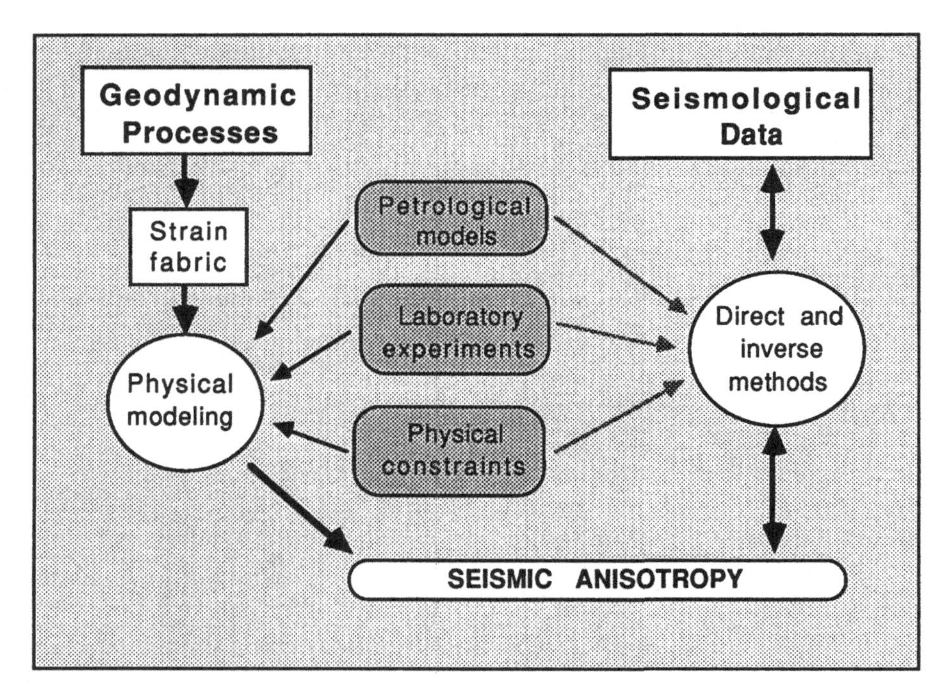

Figure 7-1 Flow chart showing the two basic approaches to investigate seismic anisotropy as a key information on the dynamics of the Earth's interior.

In the future we have to learn how to interpret the anisotropy observations with respect to deep structures and to deal with the problem of trade-off between anisotropy and heterogeneity. In particular, it is possible that in many cases the velocity anisotropy in the subcrustal lithosphere of continents could explain a large spread of velocity heterogeneities which are difficult to explain by reasonable petrological models of the upper mantle.

If we are successful in mapping the anisotropies of different scales then new anisotropic models could explain phenomena previously not understood and provide a basis for reinterpreting past seismological data such as those obtained from long-refraction profiles in the lithosphere. More realistic three-dimensional models of the velocity-depth distribution would then bring new constraints to a number of geophysical problems and provide a new insight into the Earth's past and present tectonic activity.

Glossary

The terminology used in papers related to seismic anisotropy is often confusing. We explain here several terms commonly used in seismological studies. Additional comments on this topic can be found in Grau (1988) and Crampin (1989).

Aelotropy — Anisotropy.

Anisotropy — A medium is anisotropic if its physical properties depend on the direction. In this monograph, we use the term seismic anisotropy for a large-scale elastic anisotropy (i.e. on a scale larger than a seismic wavelength).

Azimuthal anisotropy — A term used if the seismic wave properties depend on the azimuth of propagation in the horizontal plane. Structures with vertical axis of hexagonal symmetry do not exhibit azimuthal anisotropy.

Azimuthal isotropy — Hexagonal symmetry with vertical symmetry axis.

Birefringence — See shear-wave splitting.

Christoffel matrix — Symmetric real-valued matrix depending on the direction of propagation of a body wave and on the local elastic properties (its eigenvalues give the seismic velocities and its eigenvectors give the polarization directions of the waves).

Coefficient of anisotropy — Birch (1960a) defines the coefficient of anisotropy k as the percentage of velocity variation between the fastest and slowest directions, i.e.:

$$k = [(v_{max} - v_{min}) / \text{mean } v] \times 100\%,$$

where v denotes the velocity.

Compliances — Elastic coefficients relating the stress to the strain (see p. 15).

Crystallographic axis — Crystals are characterized by three axes, not necessarily mutually perpendicular, denoted [100], [010] and [001] (or a-, b- and c-axis, respectively).

Cylindrical symmetry — Symmetry around one axis (cf. hexagonal symmetry).

Elastic coefficients — Denote either the stiffnesses or the compliances of an elastic body. The number of independent elastic coefficients reduces to 1 in fluids, 2 in isotropic solids, 5 in hexagonal symmetry systems and up to 21 in the case of triclinic symmetry.

Generalized surface waves	In an anisotropic structure, the division of surface waves into two families related to their polarization, the Love and Rayleigh waves, is no longer possible. A single family of generalized surface waves has thus to be considered (Crampin, 1970).
Heterogeneity	A medium is heterogeneous if its physical properties depend on the space coordinates. A critical point is the scale of heterogeneities as compared with seismic wavelengths. For a large wavelength, an intrinsically isotropic medium with oriented heterogeneities may become equivalent to an overall anisotropic medium.
Hexagonal symmetry	Symmetry by rotation of 60° around one axis, the 6-fold symmetry axis. Hexagonal symmetry and cylindrical symmetry are equivalent for the elastic properties. There are 5 independent elastic parameters in hexagonal symmetry systems (see Table 2-2).
Hill's average	Arithmetic average of Voigt's and Reuss's averages. Voigt's and Reuss's averages correspond to the upper and lower bounds of all possible averages (Hill, 1952).
Love/Rayleigh-wave discrepancy	Observations of Love- and Rayleigh- wave dispersion curves which are not explainable by a smoothly depth-varying isotropic model.
Monoclinic symmetry	Symmetry with respect to one plane of mirror symmetry. There are 13 independent elastic parameters in the case of monoclinic symmetry.
Orthorhombic symmetry	Symmetry with respect to three mutually orthogonal planes. There are 9 independent elastic parameters for an orthorhombic material.
Orthotropy	Sometimes used for orthorhombic symmetry, in particular if one symmetry plane is horizontal.
P_H	P wave propagating horizontally (H refers here either to the horizontal direction of propagation or to the polarization direction, which are the same for P-waves).
PKIKP	P wave transmitted through the outer and inner core of the Earth.
PKiKP	P wave reflected on the surface of the inner core of the Earth.
Plane of symmetry	A plane of symmetry converts any point in the crystal into its mirror image. In crystals, planes are usually denoted

(n,m,p) where n, m, p are Miller's indices, e.g. plane (010) is the plane containing axes [100] and [001].

Polarization anisotropy — A term sometimes used by seismologists in contrast to azimuthal anisotropy to refer to an anisotropic medium in which the seismic-wave properties depend on the polarization of the waves, and not on the azimuth of their propagation direction. S_H / S_V differences of velocity or Love/Rayleigh-wave discrepancies are often considered as due to "polarization anisotropy" while azimuthal variations of velocity would be due to "azimuthal anisotropy". As both types of behavior may occur simultaneously, this distinction is rather artificial and may cause confusion (Mitchell, 1984). A clear mathematical separation of the two types of anisotropy for surface waves is given by Montagner and Nataf (1986).

P_V — P wave propagating along the vertical direction (see comments on P_H).

***Quasi* P or *Quasi* S waves** — Terms sometimes used in a weakly anisotropic medium to characterize the seismic waves whose directions of polarization are either close to the propagation direction (*Quasi* P) or to the normal to this direction (*Quasi* S).

***Quasi* Love or *Quasi* Rayleigh waves** — Terms used for Love or Rayleigh waves propagating in a weakly anisotropic medium, where the polarization anomalies are small.

Radial anisotropy — A term sometimes used to denote azimuthal isotropy (or hexagonal symmetry with a vertical symmetry axis - see also transverse isotropy).

Reuss's average — Volume average of the elastic compliances in aggregates based on the assumption of a homogeneous stress (Reuss, 1929). Reuss's averages K^R and G^R are:

$K^R = [3(a+2b)]^{-1}$ (bulk modulus),
$G^R = 5/(4a-4b+3c)$ (shear modulus), where
$3a = S_{11} + S_{22} + S_{33}$,
$3b = S_{23} + S_{31} + S_{12}$,
$3c = S_{44} + S_{55} + S_{66}$,

and where S_{nm} are the compliances in matrix notations (see section 2.2)

ScS — Shear wave reflected at the core-mantle boundary.

S_H, S_V — S-waves with polarization direction in the horizontal and vertical planes, respectively. In general anisotropic media, S-wave splitting occurs and, except in particular symmetry situations and/or propagation directions, S_H and S_V notations are meaningless.

Shear-wave splitting — Behavior of S waves in an anisotropic medium which are split into two shear waves with mutually perpendicular polarization directions. By analogy to optics, shear-wave splitting is sometimes called birefringence.

SKS — Mantle shear wave transmitted through the outer core as a P wave (see p. 88).

Stiffnesses — Elastic coefficients relating the strain to the stress such as rigidity, bulk modulus, or tensor components c_{ijkl} (see section 2-2)

Transverse isotropy — A term introduced by Love (1927) to describe a medium with one axis of cylindrical symmetry, which is for the elastic-wave propagation equivalent to hexagonal symmetry. In the geophysical literature, this axis is often implicitly considered to be vertical, so that velocities and polarizations in transversely isotropic structures are independent of the azimuth of propagation in the horizontal plane.

Velocity anisotropy — Terms used to denote the directional dependence of velocity. Percentage of maximum velocity variations is often called coefficient of anisotropy.

Voigt's average — Volume average of the elastic stiffnesses in aggregates based on the assumption of a homogeneous strain (Voigt, 1928). Voigt's averages K^V and G^V are:

$K^V = (A + 2B) / 3$ (bulk modulus),
$G^V = (A - B + 3C) / 5$, (shear modulus), where
$3A = C_{11} + C_{22} + C_{33}$,
$3B = C_{23} + C_{31} + C_{12}$,
$3C = C_{44} + C_{55} + C_{66}$,

and where C_{nm} are the stiffnesses in matrix notation (see section 2.2).

REFERENCES

Achauer, U., Glahn, A. Granet, M., Wittlinger, G. and Slack, P.D., 1989. P-Delay-Time Tomography of the Lithosphere-Asthenosphere System Beneath the Rhinegraben Rift, *EOS, 70,* 1221, (Abstract).

Aki, K. and Kaminuma, K., 1963. Phase velocity in Japan. Part. I. Love waves from the Aleutian shock of March 9, 1957, *Bull. Earthq. Res. Inst., 41*, 243-259.

Aki, K. and Richards, P.G., 1980. *Quantitative seismology: theory and methods,* Freeman & Co., San Francisco, vol. I, 932 pp.

Aki, K., Christoffersson, A. and Husebye, E.S., 1977. Determination of the three-dimensional seismic structure of the lithosphere, *J. Geophys. Res., 82*, 277-296.

Aleksandrov, K.S. and Ryzhova, T.V., 1961a. The elastic properties of crystals, *Sov. Phys. Crystallogr., 6,* 228-252.

Aleksandrov, K.S. and Ryzhova, T.V., 1961b. The elastic properties of rock-forming minerals, I: pyroxenes and amphiboles, *Izv. Acad. Sci. USSR, Geophys. Ser., 9*, 1339-1344, (in Russian).

Aleksandrov, K.S. and Ryzhova, T.V., 1961c. The elastic properties of rock-forming minerals,II: layered silicates, *Izv. Acad. Sci. USSR, Geophys. Ser., 12,* 1799-1804, (in Russian).

Aleksandrov, K.S. and Ryzhova, T.V., 1961d. Moduli of elasticity of pyrite, *Izv. Acad. Sci. USSR, Sibir. Branch, 6*, 43-47, (in Russian).

Aleksandrov, K.S. and Ryzhova, T.V., 1962. The elastic properties of rock-forming minerals, III: feldspars, *Izv. Acad. Sci. USSR, Geophys. Ser., 2*, 186-189, (in Russian).

Aleksandrov, K.S., Ryzhova, T.V. and Belikov, B.P., 1963. The elastic properties of pyroxenes, *Sov. Phys. Crystallogr., 8*, 738-741, (in Russian).

Aleksandrov, K.S., Alchikov, V.V., Belikov, B.P., Zaslavskii, B.I. and Krupnyi, A.I., 1974. Velocities of elastic waves in minerals at atmospheric pressure and increasing of precision of elastic constants by means of EVM, *Izv.Acad. Sci. USSR, Geol.Ser., 10,* 15-24, (in Russian).

Allard, B. and Sotin, C., 1989. Determination of percentages of different mineral phases in granular rocks using image analysis on a microcomputer, *Computers & Geosciences, 15*, 441-448.

Anderson, D.L., 1961. Elastic wave propagation in layered anisotropic media, *J. Geophys. Res.*, 82, 277-296.

Anderson, D.L., 1962. Love wave dispersion in heterogeneous anisotropic media, *Geophysics*, 27, 445-454.

Anderson, D.L., 1966. Recent evidence concerning the structure and composition of the Earth's mantle, *Phys. Chem. Earth, 6,* 1-131.

Anderson, D.L., 1987. Thermally induced phase changes, lateral heterogeneities of the mantle, continental roots and deep slab anomalies, *J. Geophys. Res., 92,* 13968-13980.

Anderson, D.L., 1989. *Theory of the Earth,* Blackwell Sci. Publ., 366 pp.

Anderson, D.L. and Archambeau, C.B., 1964. The anelasticity of the Earth, *J.Geophys. Res., 69,* 2071-2084.

Anderson, D.L. and Harkrider, D., 1962. The effect of anisotropy on continental and oceanic surface wave dispersion, *J. Geophys. Res., 67,* 1627, (Abstract).

Anderson, D.L. and Regan, J., 1983. Upper mantle anisotropy and the oceanic lithosphere, *Geophys. Res. Lett., 10,* 841-844.

Anderson, D.L. and Sammis, C., 1970. Partial melting in the upper mantle, *Phys. Earth Planet. Int., 3,* 41-50.

Anderson, D.L., Minster, B. and Cole, D., 1974. The effect of oriented cracks on seismic velocities, *J. Geophys. Res., 79,* 4011-4015.

Anderson, O.L., 1986. Properties of iron at the Earth's core conditions, *Geophys. J. R. astr. Soc., 84,* 561-579.

Anderson, O.L. and Liebermann, R.C., 1966. Sound velocities in rocks and minerals,*VESIAC State-of-the-Art Report,* Univ. of Michigan, Ann Arbor.

Ando, M., 1984. ScS polarization anisotropy around the Pacific Ocean, *J. Phys. Earth, 32,* 179-195.

Ando, M., Ishikawa, Y. and Yamazaki, F., 1983. Shear wave polarization anisotropy in the mantle beneath Honshu, Japan, *J. Geophys. Res., 88,* 5850-5864.

Ansel, V. and Nataf, H.C., 1989. Anisotropy beneath 9 stations of the Geoscope broadband network as deduced from shear-wave splitting, *Geophys. Res. Lett., 16,* 409-412.

Ansorge, J., Bonjer, K.-P. and Emter, D., 1979. Structure of the uppermost mantle from long-range seismic observations in Southern Germany and the Rhinegraben area, *Tectonophys., 56,* 31-48.

Asada, T., 1984. Seismic anisotropy beneath the ocean, *J. Phys. Earth, 32,* 177-178.

Aster, R.C., Shearer, P.M. and Berger, J., 1990. Quantitative measurements of shear wave polarizations at the Anza seismic network, southern California: Implications for shear wave splitting and earthquake prediction, *J. Geophys. Res., 95,* 12449-12473.

Ave Lallemant, H.G. and Carter, N.L., 1970. Syntectonic recrystallization of olivine and modes of flow in the upper mantle, *Geol. Soc. Am. Bull., 81,* 2203-2220.

Babuska, V., 1968. Elastic anisotropy of igneous and metamorphic rocks, *Stud. Geophys. Geod. 12,* 291-303.

Babuska, V., 1972a. Anisotropy of the upper mantle rocks., *Z. Geophys., 38,* 461-467.

Babuska, V., 1972b. Elasticity and anisotropy of dunite and bronzitite, *J. Geophys. Res., 77,* 6955-6965,

Babuska, V., 1976. Elastic properties of ultrabasic xenoliths at pressures to 10 kilobars, *Izv. Akad. Nauk USSR, Fiz. Zemli, 6,* 22-33 (in Russian).

Babuska, V., 1981. Anisotropy of V_P and V_S in rock-forming minerals, *J. Geophys., 50,* 1-6.

Babuska, V., 1984. P-wave velocity anisotropy in crystalline rocks, *Geophys. J.R.astr. Soc., 76,* 113-119.

Babuska, V. and Gueguen, Y., 1989. Anisotropy in the deep lithosphere and long seismic profiles, *Terra Nova, 1,* 91, (Abstract).

Babuska, V. and Plomerova, J., 1989. Seismic anisotropy of the subcrustal lithosphere in Europe: Another clue to recognition of accreted terranes? in *"Deep structure and past kinematics of accreted terranes", J.W. Hillhouse (ed.), Geophys. Monograph, 50, AGU,* 209-217.

Babuska, V. and Plomerova, J., 1991. The lithosphere in central Europe - seismological and petrological aspects, *Tectonophys.,* in press.

Babuska, V. and Pros, Z., 1984. Velocity anisotropy in granodiorite and quartzite due to the distribution of microcracks, *Geophys. J.R. astr. Soc., 76,* 121-127.

Babuska, V. and Sileni, J., 1981. Velocity anisotropy and symmetry of dunite fabric, *Studia Geoph. Geod., 25*, 231-244.

Babuska, V., Pros, Z. and Franke, W., 1977. Effect of fabric and cracks on the elastic anisotropy in granodiorite, *Publs. Inst. Geophys. Pol. Acad. Sci. A-6, 117*, 179-186.

Babuska, V., Fiala, J., Kumazava, M., Ohno, I. and Sumino, Y., 1978a. Elastic properties of garnet solid-solution series, *Phys. Earth Planet. Int., 16*, 157-176.

Babuska, V., Fiala, J., Mayson, D.J. and Liebermann, R.C., 1978b. Elastic properties of eclogite rocks from the Bohemian Massif., *Studia Geophys. Geod., 22*, 348-361.

Babuska, V., Plomerova, J. and Sileni, J., 1984. Large-scale oriented structures in the subcrustal lithosphere of Central Europe, *Ann. Geophys., 2*, 649-662.

Babuska, V., Plomerova, J. and Sileni, J., 1987. Structural model of the subcrustal lithosphere in central Europe, in *The composition, structure and dynamics of the lithosphere-asthenosphere system, C. Froidevaux and K. Fuchs (eds.), AGU Geodyn. Series, 16*, 239-251.

Babuska, V., Plomerova, J. and Sileni, J., 1991. Model of seismic anisotropy in deep lithosphere of central Europe, submitted to *Phys. Earth Planet. Int.*.

Backus, G.E., 1962. Long-wave elastic anisotropy produced by horizontal layering, *J. Geophys. Res., 67*, 4427-4440.

Backus, G.E., 1965. Possible forms of seismic anisotropy of the uppermost mantle under oceans, *J. Geophys. Res., 70*, 3429-3439.

Backus, G.E., 1967. Converting vector and tensor equations to scalar equations in spherical coordinates, *Geophys. J. R. astr. Soc., 13*, 71-101.

Backus, G.E. and Gilbert, J.F., 1968. The resolving power of gross earth data, *Geophys.J. R. astr. Soc., 16*, 169-205.

Backus, G.E. and Gilbert, J.F., 1970. Uniqueness in the inversion of inaccurate gross earth data, *Phil. Trans. R. Soc. London, 266*, 123-192.

Ballard, R.D. and van Andel, T.H., 1977. Morphology and tectonic of the inner rift valley at lat. 36.5° N on the Mid-Atlantic Ridge., *Geol. Soc. Am. Bull., 88*, 507-530.

Bamford, D., 1973. Refraction data in Western Germany - a time-term interpretation, *Z. Geophys., 39*, 907-927.

Bamford, D., 1977. Pn velocity anisotropy in a continental upper mantle, *Geophys. J. R. astr. Soc., 49*, 29-48.

Bamford, D., Jentsch, M. and Prodehl, C., 1979. Pn anisotropy studies in northern Britain and the eastern and western United States, *Geophys. J. R. astr. Soc., 57*, 397-429.

Barrel, J., 1914. The strength of the Earth's crust, *J. Geol., 22*, 23.

Bass, J.D. and Anderson, D.L., 1984. Composition of the upper mantle: geophysical tests of two petrological models, *Geophys. Res. Lett., 11*, 229-232.

Bayuk, E.I., 1966. Velocities of elastic waves in samples of metamorphic rocks at pressures to 4 kilobars, in *Electrical and Mechanical Properties of Rocks under High Pressures*, Nauka, Moscow, 16-36, (in Russian).

Bayuk, E.I., Volarovich, A.P., Klima, K., Pros, Z. and Vanek, J., 1967. Velocity of longitudinal waves in eclogite and ultrabasic rocks under pressure to 4 kilobars, *Studia Geoph. Geod., 11*, 271-280.

Bayuk, E.I., Volarovich, M.P. and Skvortsova, L.S., 1971. Velocities of elastic waves measured at high pressures in igneous and metamorphic rocks of various regions, in *Tectonophysical and Mechanical Properties of Rocks*, Nauka, Moscow, 127-137 (in Russian).

Bayuk, E.I., Volarovich, M.P. and Levitova, F.M., 1982. *Elastic Anisotropy of Rocks under High Pressure,* Nauka, Moscow, (in Russian).
Bean, C.J. and Jacob, A.W.B., 1990. P-wave anisotropy in the lower lithosphere, *Earth Planet. Sci. Lett., 99,* 58-65.
Belikov, B.P., Aleksandrov, K.S. and Ryzhova, T.V., 1970. *Elastic properties of rock-forming minerals and rocks,* Nauka, Moscow, (in Russian).
Ben-Avraham, Z., Nur, A. and Jones, D., 1982. The emplacement of ophiolites by collision, *J. Geophys. Res., 87,* 3861-3867.
Berge, P.A., Mallick, S., Fryer, G.J., Barstov, N., Carter, J.A., Sutton, G.H. and Ewing, J.I., 1991. In situ measurements of transverse isotropy in shallow-water marine sediments, *Geophys. J. Int., 104,* 241-254.
Bernal, J.D., 1936. Hypothesis on 20° discontinuity, *Observatory, 59,* 268.
Bibee, L.D. and Shor, G.G., 1976. Compressional wave anisotropy in the crust and upper mantle, *Geophys. Res. Lett., 3,* 639-642.
Birch, F., 1960a. The velocity of compressional waves in rocks to 10 kilobars, *J. Geophys. Res., 65,* 1083-1102.
Birch, F., 1960b. Elastic constants of rutile - A correction to a paper by R.K. Verma, Elasticity of some high-density crystals, *J. Geophys. Res., 65,* 3855-3856.
Birch, F., 1961. The velocity of compressional waves in rocks to 10 kilobars: Part 2, *J. Geophys. Res., 66,* 2199-2224
Birch, F., 1966. Compressibility; elastic constants, in *Handbook of Physical Constants, revised edition, Geol. Soc. Am. Mem., 97, S.P. Clark, Jr., ed.,* 97-173, New York.
Blankenship, D.D., Alley R.B. and Bentley, C.R., 1989. Fabric development in ice sheets: Seismic anisotropy, *EOS, 70,* 462, (Abstract).
Blenkinsop, T.G., 1988. Fracture anisotropy in a fold and thrust belt: field observations, *The Third Int. Workshop on Seismic Anisotropy,* Berkeley, (Abstract).
Blenkinsop, T.G., 1990. Correlation of paleotectonic fracture and microfracture orientations in cores with seismic anisotropy of Cajon Pass drill hole, southern California, *J. Geophys. Res., 95,* 143-150.
Bolt, B.A. and Nuttli, O.W., 1966. P wave residuals as a function of azimuth, 1, observations, *J. Geophys. Res., 71,* 5977-5985.
Boore, D.M., 1969. Effect of higher mode contamination on measured Love-wave phase velocities, *J. Geophys. Res., 74,* 6612-6616.
Booth, D.C. and Crampin, S., 1983. The anisotropic reflectivity technique: theory, *Geophys. J. R. astr. Soc., 72,* 755-766.
Booth, D.C. and Crampin, S., 1985. Shear-wave polarizations on a curved wavefront at an isotropic free surface, *Geophys. J. R. astr. Soc., 83,* 31-45.
Borevsky, L.V., Vartanyan, G.S. and Kulilov, T.B., 1987. Hydrogeological essay, in *"The superdeep well of the Kola Penninsula, Y.A. Kozlovsky (ed.), Springer-Verlag,* New York, 271-287.
Brace, W.F., 1965. Some new measurements of linear compressibility of rocks, *J. Geophys. Res., 70,* 391-398.
Brace, W.F., Paulding, B.W., Jr. and Scholz, C., 1966. Dilatancy in the fracture of crystalline rocks, *J. Geophys. Res., 71,* 3939-3953.
Brocher, T.M., Ambos, E.L., Fisher, M.A., Fuis, G.S., Luetgert, J.H., McCarthy, J. and Mooney, W.D., 1988. Seismic anisotropy observations in continental crust based on recent USGS seismic refraction experiments, *The Third Int. Workshop on Seismic Anisotropy,* Berkeley, (Abstract).
Brodov, L.Y., Evstifeyev, V.I., Karus, E.V. and Kulichikhina, T.N., 1984. Some

results of the experimental study of seismic anisotropy of sedimentary rocks using different types of waves, *Geophys. J. R. astr. Soc.*, *76*, 191-200.
Buchwald, V.T., 1961. Rayleigh waves in anisotropic media, *Q. J. Mechs. appl. Math.*, *14*, 461-469.
Buchwald, V.T. and Davis, A., 1963. Surface waves in elastic media with cubic symmetry, *Q. J. Mechs. appl. Math.*, *16*, 283-294.
Bullen, K.E. and Bolt, B., 1985. *An introduction to the theory of seismology*, Cambridge Univ. Press, Cambridge, 499 pp.
Burdock, J.H., 1980. Seismic velocity structure of the metamorphic belt of central California, *Bull. Seismol. Soc. Am.*, *70*, 203-221.
Caloi, P., 1954. L'astenosfera come canale-guida dell'energia sismica, *Annali Geofis.*, *7*, 491.
Cann, J.R., 1974. A model for oceanic crustal structure developed, *Geophys. J. R. astr. Soc.*, *39*, 169-187.
Cara, M., 1978. Regional variations of higher Rayleigh-mode phase velocities: a spatial-filtering method, *Geophys. J. R. astr. Soc.*, *54*, 439-460.
Cara, M., 1979. Lateral variations of S-velocity in the upper mantle from higher Rayleigh modes, *Geophys. J. R. astr. Soc.*, *57*, 649-670.
Cara, M., 1983. Crust-mantle structure inferred from surface-waves, in *"Earthquakes: Observation, Theory and Interpretation", H. Kanamori and E. Boschi (eds.)*, North-Holland Publ. Co., 319-323.
Cara, M. and Leveque, J.J., 1987. Oriented olivine crystals in the upper mantle: a test from the inversion of multimode surface-wave data, *Phys. Earth Planet. Inter.*, *47*, 246-252.
Cara, M. and Leveque, J.J., 1988. Anisotropy of the asthenosphere: the higher mode data of the Pacific revisited, *Geophys. Res. Lett.*, *15*, 205-208.
Cara, M., Nercessian, A. and Nolet, G., 1980. New inferences from higher mode data in western Europe and northern Eurasia, *Geophys. J. R. astr. Soc.*, *61*, 459-478.
Cara, M., Leveque, J.J. and Maupin, V., 1984. Density-versus-depth models from multimode surface waves, *Geophys. Res. Lett.*, *11*, 633-636.
Carlson, R.L., Schaftenaar, C.H. and Moore, R.P., 1984. Causes of compressional-wave anisotropy in carbonate-bearing, deep-sea sediments, *Geophysics*, *49*, 525-532.
Carmichael, R.S., 1984. *Handbook of physical properties of rocks*, *CRC Press, vol. 3*, Boca Raton, Florida, 340 pp.
Carter, N.L. and Tsenn, M.C., 1987. Flow properties of the continental lithosphere, *Tectonophys.*, *136*, 27-63.
Carter, N.L., Baker, D.W. and George R.P., Jr., 1972. Seismic anisotropy, flow and constitution of the upper mantle, in *"Flow and fracture of rocks", A.G.U. Geophys. Monograph 16, H.C. Heard, I.Y. Borg, N.L. Carter, C.B. Raleigh, (eds.)*, Washington D.C., 167-190.
Cerveny, V., 1972. Seismic rays and ray intensities in inhomogeneous anisotropic media, *Geophys. J. R. astr. Soc.*, *29*, 1-13.
Cerveny, V., 1989. Ray tracing in factorized anisotropic media, *Geophys. J. Int.*, *99*, 91-100.
Cerveny, V., Molotkov, I.A. and Psencik, I., 1977. *Ray method in seismology*, Charles Univ. Press, Prague.
Chan, W.W. and Silver, P.G., 1990. Seismic anisotropy in China from SKS, *EOS*, *71*, 555, (Abstract).
Chen, K.H., 1984. Numerical modeling of elastic wave propagation in anisotropic heterogeneous media: a finite element approach, *SEG Convention Expanded*

Abstracts, 631-634.
Chopin, Ch., 1987. Very high pressure metamorphism in the western Alps: new petrologic and field data, *Terra Cognita, 7,* 94, (Abstract).
Christensen, N.I., 1965. Compressional wave velocities in metamorphic rocks at pressures to 10 kilobars, *J. Geophys. Res., 70,* 6147-6164.
Christensen, N.I., 1966a. Shear wave velocities in metamorphic rocks at pressures to 10 kilobars. *J. Geophys. Res., 71,* 3549-3556.
Christensen, N.I., 1966b. Elasticity of ultrabasic rocks, *J. Geophys. Res., 71,* 5921-5932.
Christensen, N.I., 1978. Ophiolities, seismic velocities and oceanic crustal structure, *Tectonophys., 47,* 131-157.
Christensen, N.I., 1984. The magnitude, symmetry and origin of upper mantle anisotropy based on fabric analyses of ultramafic tectonites, *Geophys. J.R. astr.Soc., 76,* 89-111.
Christensen, N.I., Crosson, R.S., 1968. Seismic anisotropy in the upper mantle, *Tectonophys., 6,* 93-102.
Christensen, N.I. and Fountain, D.M., 1975. Constitution of the lower continental crust based on experimental studies of seismic velocities in granulite, *Bull. Geol. Soc. Am., 86,* 227-236.
Christensen, N.I. and Ramananantoandro, R., 1971. Elastic moduli and anisotropy of dunite to 10 kilobars, *J. Geophys. Res., 76,* 4003-4021.
Christensen, N.I. and Salisbury, M.H., 1979. Seismic anisotropy in the oceanic upper mantle: evidence from the Bay Islands Ophiolite Complex, *J. Geophys. Res., 84,* 4601-4610.
Christensen, N.I. and Szymanski, D.L., 1988. Origin of reflections from the Breward fault zone, *J. Geophys. Res., 93,* 1087-1102.
Christensen, U.R., 1987. Some geodynamical effects of anisotropic viscosity, *Geophys. J. R. astr. Soc., 91,* 711-736.
Cleary, J., 1967. Azimuthal variation of the Lohgshot source term, *Earth Planet. Sci. Lett., 3,* 29-37.
Cleary, J. and Hales, A.L., 1966. An analysis of the travel times of P waves to North American stations in the distance range 32 to 100, *Bull. Seismol. Soc. Am., 56,* 467-489.
Coates, R.T. and Chapman, C.H., 1990. Quasi-shear wave coupling in weakly anisotropic 3-D media, *Geophys. J. Int., 103,* 301-320.
Cook, F.A., 1986. Contineta1 evolution by lithosphere shingling, in *"Reflection Seismology: The Continental Crust", M. Barazangi and L. Brown (eds.), Geodyn. Ser., 14, AGU,* Washington D.C., 13-19.
Cormier, V.F., 1986. Synthesis of body waves in transversely isotropic earth models, *Bull. Seismol. Soc. Am., 76,* 231-240.
Coyner, K.B., 1988. Laboratory measurements of elastic anisotropy in dry and saturated Chelmsford granite, *The Third Int. Workshop on Seismic anisotropy,* Berkeley, USA, (Abstract).
Crampin, S., 1970. The dispersion of surface waves in multilayered anisotropic media, *Geophys. J. R.astr. Soc., 21,* 387-402.
Crampin, S., 1975. Distinctive particle motion of surface waves as a diagnostic of anisotropic layering, *Geophys. J. R. astr. Soc., 40,* 177-186.
Crampin, S., 1977. A review of the effects of anisotropic layering on the propagation of seismic waves, *Geophys. J.R. astr. Soc., 49,* 9-27.
Crampin, S., 1978. Sesmic-wave propagation through a cracked solid: Polarization as a possible dilatancy diagnostic, *Geophys. J.R. astr. Soc., 53,* 467-496.

Crampin, S., 1982. Comments on "Possible forms of anisotropy of the uppermost mantle under oceans" by George E. Backus, *J. Geophys. Res.*, *87*, 4636-4640.

Crampin, S., 1984a. Effective anisotropic elastic constants for wave propagation through cracked solids, *Geophys. J. R. astr. Soc.*, *76*, 135-145.

Crampin, S., 1984b. Anisotropy in exploration geophysics, *First Break*, *2*, 19-21.

Crampin, S., 1984c. An introduction to wave propagation in anisotropic media, *Geophys. J. R. astr. Soc.*, *76*, 135-145.

Crampin, S., 1985. Evidence for aligned cracks in the Earth's crust, *First Break*, *3*, 12-15.

Crampin, S., 1987. Geological and industrial implications of extensive-dilatancy anisotropy, *Nature*, *328*, 491-496.

Crampin, S., 1989. Suggestions for a consistent terminology for seismic anisotropy, *Geophys. Prosp.*, *31*, 753-770.

Crampin, S. and Booth, D.C., 1985. Shear-wave polarizations near the North Anatolian fault, II, Interpretation in terms of crack-induced anisotropy, *Geophys. J.R. astr. Soc.*, *83*, 75-92.

Crampin, S. and King, D.W., 1977. Evidence for anisotropy in the upper mantle beneath Eurasia from the polarization of higher mode seismic surface waves, *Geophys. J. R. astr. Soc.*, *49*, 59-85.

Crampin, S., Evans, R., Ucer, B., Doyle, M., Davis, J.P, Yegorkina, G.V. and Miller, A., 1980. Observations of dilatancy induced polarization anomalies and earthquake prediction, *Nature*, *286*, 874-877.

Crampin, S., Chesnokov, E.M. and Hipkin, R.G., 1984a. Seismic anisotropy - the state of the art: II, *Geophys. J.R. astr. Soc.*, *76*, 1-16.

Crampin, S., Evans, R. and Atkinson, B.K., 1984b. Earthquake prediction: A new physical basis, *Geophys. J.R. astr. Soc.*, *76*, 147-156.

Crampin, S., Booth, D.C., Evans, R., Peacock, S. and Fletcher, J., 1990. Changes in shear wave splitting at Anza near the time of the North Palm Springs earthquake, *J. Geophys. Res.*, *95*, 11197-11212.

Crampin, S., Booth, D., Evans R., Peacock, S. and Fletcher, J.B., 1991. Response to comments by Aster et al. (1990) about temporal variations in shear-wave splitting observed by Peacock et al. (1988) and Crampin et al. (1990), *J. Geophys. Res.*, *96*, 6403-6414.

Dahlen, F.A., 1972. Elastic velocity anisotropy in the presence of an anisotropic initial stress, *Bull. Seismol. Soc. Am.*, *62*, 1183-1193.

Daley, P.F. and Hron, F., 1977. Reflection and transmission coefficients for transversely isotropic media, *Bull. Seismol. Soc. Am.*, *67*, 661-675.

Daley, P.F. and Hron, F., 1979. Reflection and transmission coefficients for seismic waves in elliptically anisotropic media, *Geophysics*, *44*, 23-38.

Daley, T.M., McEvilly , T.V. and Majer E.L., 1988. Multiply-polarized shear-wave VSPs from the Cajon Pass drillhole, *Geophys. Res. Lett.*, *15*, 1001-1004.

Daley T.M. and McEvilly, T.V., 1990. Shear wave anisotropy in the Parkfield Varian well VSP, *Bull. Seismol. Soc. Am.*, *80*, 857-869.

Daly, 1940. *Strength and structure of the Earth*, Prentice-Hall, 434 pp.

Deer, W.A., Howie, R.A. and Zussman, J ., 1966. *An introduction to the rockforming minerals, London: Longmans*, Green and Co. Ltd.

Dell'Angelo, L. and Tullis, J., 1989. Fabric development in experimentally sheared quartzites, *Tectonophys.*, *169*, 1-21.

Dick, H.J.B., 1987. Petrologic variability of the oceanic uppermost mantle, in *Geophysics and petrology of the deep crust and upper mantle, Noller, J.S., Kirby, S.H. and Nielson-Pike, J.E., (eds.), U.S. Geol. Survey Circular 956,*

17-20.

Doornbos, D.J., Spilopoulos, S. and Stacey, F.D., 1985. Seismological properties of D" and the structure of the thermal boundary layer, *Phys. Earth Planet. Int., 41,* 225-239.

Dorman, J., Ewing, M. and Oliver, J., 1960. Study of shear velocity distribution in the upper mantle by mantle Rayleigh waves, *Bull. Seismol. Soc. Am., 50,* 87-115.

Dost, B., 1990. Upper mantle structure under western Europe from fundamental mode and higher mode surface waves using the NARS array, *Geophys. J. Int., 100,* 131-151.

Duffy, T.S. and Anderson, D.L., 1988. Seismic velocities in mantle minerals and the mineralogy of the upper mantle, *J. Geophys., Res., 94,* 1895-1912.

Dziewonski, A.M. and Anderson, D.L., 1981. Preliminary reference earth model, *Phys. Earth Planet. Int., 25* , 297-356.

Dziewonski, A. M. and Anderson, D.L., 1983. Travel times and station corrections for P waves at teleseismic distances, *J. Geophys. Res., 88,* 3295-3314.

Evans, R., 1984a. Anisotropy : a pervasive feature of fault zones?, *Geophys. J.R. astr. Soc., 76,* 157-163.

Evans, R., 1984b. Effects of the free surface on shear waves, *Geophys. J. R. astr. Soc., 76,* 165-172.

Estey, L.H. and Douglas, B.J., 1986. Upper-mantle anisotropy: a preliminary model, *J. Geophys. Res., 91,* 11393-11406.

Farra, V., 1989. Ray perturbation theory for heterogeneous hexagonal anisotropic media, *Geophys. J. Int., 99,* 723-737.

Farra, V., 1990. Amplitude computation in heterogeneous media by ray perturbation theory: a finite element approach, *Geophys. J. Int., 103,* 341-354.

Fedorov, F.I., 1968. Theory of elastic waves in crystals, *Plenum Press,* New York.

Finlayson, D.M., Leven, J.H. and Wake-Dyster, K.D., 1989. Large-scale lenticles in the lower crust under an intra-continental basin in eastern Australia., in *"Properties and Processes of Earth's Lower Crust", R.F. Mereu et al. (eds.), Geophys. Monograph, AGU, 51,* 3-16.

Forsyth, D.W., 1975. The early structural evolution and anisotropy of the oceanic upper mantle, *Geophys. J. R. astr. Soc., 43,* 103 - 162.

Fountain, D.M., 1987. Geological and geophysical nature of the lower continental crust as revealed by exposed cross sections of the continental crust, in *"Geophysics and Petrology of the Deep Crust and Upper Mantle", Noller, J.S., Kirby, S.H. and Nielson-Pike, J.E., (eds.), U.S. Geol. Survey Circular 956,* 25-26.

Fountain, M., Hurich, C.A. and Smithson, S.B., 1984. Seismic reflectivity of mylonite zones in the crust, *Geology, 12,* 195-198.

Francis, T.J.G., 1969. Generation of seismic anisotropy in the upper mantle along the mid-oceanic ridges, *Nature, 221,* 162-165.

Francis, T.G.J., 1976. The ratio of compressional to shear velocity and rock porosity on the axis of the Mid-Atlantic Ridge, *J. Geophys. Res., 81,* 4361-4364.

Franklin, J.N., 1970. Well-posed stochastic extensions of ill-posed linear problems, *J. Math. Anal. Appl., 31,* 682-716.

Frazer, L.N. and Fryer, G.J., 1989. Useful properties of the system matrix for a homogeneous anisotropic visco-elastic solid, *Geophys. J. Int., 97,* 173-177.

Fryer, G.J., 1988. Transverse isotropy in the upper oceanic crust, *The Third Int. Workshop on Seismic Anisotropy,* Berkeley, (Abstract).

Fryer, G.J. and Frazer, L. N., 1984. Seismic waves in stratified anisotropic media, *Geophys. J. R. astr. Soc., 78,* 691-710.

Fuchs, K., 1977. Seismic anisotropy of the subcrustal lithosphere as evidence for dynamical processes in the upper mantle, *Geophys. J. R. astr. Soc., 49*, 167-179.

Fuchs, K., 1983. Recently formed elastic anisotropy and petrological models for the continental subcrustal lithosphere in southern Germany, *Phys. Earth Planet. Int., 31*, 93-118.

Fuchs, K. and Vinnik, L.P., 1982. Investigation of the subcrustal lithosphere and asthenosphere by controlled source seismic experiments on long range profiles, in *"Continental and oceanic rifts", Palmason, G. (ed.), Geodyn Ser. AGU-GSA*, Washington D.C., 81-89.

Fuchs, K., Vinnik, L.P. and Prodehl, C., 1987. Exploring heterogeneities of the continental mantle by high resolution seismic experiments, in *"Composition, Structure and Dynamics of the Lithosphere-Asthenosphere System", Fuchs, K. and Froidevaux, C. (eds.), AGU*, Washington, D.C., 137-154.

Fukao, Y., 1984. ScS evidence for anisotropy in the earth's mantle, *Nature, 309*, 695-698.

Furnish, M.D. and Basset, W.A., 1983. Investigation of the mechanism of the olivine-spinel transition in fayalite by synchrotron radiation, *J. Geophys. Res., 88*, 10333-10341.

Gaiser, J.E. and Carrigan, D., 1990. Observation of azimuthal anisotropy in the near-surface, *The Fourth Int. Workshop on Seismic Anisotropy*, Edinburgh, (Abstract).

Gajewski, D. and Prodehl, C., 1987. Seismic refraction investigation of the Black Forest, *Tectonophys., 142*, 27-48.

Gajewski, D. and Psencik, I., 1987. Computation of high-frequency wavefield in 3-D laterally inhomogeneous anisotropic media, *Geophys. J. R. astr. Soc., 91*, 383-411.

Gajewski, D. and Psencik, I., 1990. Vertical seismic profile synthetics by dynamic ray tracing in laterally varying layered anisotropic structures, *J. Geophys. Res., 95*, 11301-11315.

Galdin, N.E., 1977. Physical properties of igneous and metamorphic rocks at high pressure and temperature, Nauka, Moscow (in Russian).

Garbin, H.D. and Knopoff, L., 1975. Elastic moduli of a medium with liquid-filled cracks, *Q. appl. Math., 33*, 301-303.

Gariel, G.C., Archuleta, R.J. and Bouchon, M., 1990. Rupture process of an earthquake with kilometric size fault inferred from the modeling of near source records, *Bull. Seismol. Soc. Am.*, 80, 870-888.

Garmany, J., 1983. Some properties of elastodynamic eigensolutions in stratified media, *Geophys. J. R. astr. Soc., 75*, 565-569.

Garmany, J., 1988a. Seismograms in stratified anisotropic media - I. WKBJ theory, *Geophys. J. R. astr. Soc., 92*, 365-377.

Garmany, J., 1988b. Seismograms in stratified anisotropic media - II. Uniformly asymptotic approximations, *Geophys. J. R. astr. Soc., 92*, 379-389.

Garmany, J., 1989. A student garden of anisotropy, *Ann. Rev. Earth Planet. Sci., 17*, 285-308.

Gebrande, H., 1982. Elastic wave velocities and constans of elasticity of rocks and rock-forming minerals, in *Landolt-Bornstein, Numerical Data and Functional Relationships in Science and Technology, v. 1, Phys. Properties of Rocks, Springer-Verlag, K.H. Hellwege (ed.)*', Berlin, Heidelberg, New York, 1-99.

Gelfand, I.M. and Shapiro, Z.Y., 1956. Representation of the group of rotations in three-dimensional space and their applications, *Am. Math. Soc. Trans.*, 2, 207-316.

Giardini, D., Xiang-Dong L. and Woodhouse, J.H., 1987. Three-dimensional structure of the Earth from splitting in free-oscillation spectra, *Nature, 325,* 405-411.

Glahn, A. and Granet, M., 1991. 3-D P-wave velocity structure of the southern Rhine Graben area and implications on the rift's dynamic evolution, submitted to *Geophys. J. Int.*.

Gledhill, K.R., 1990. Evidence for shallow crustal anisotropy in the Wellington region, New Zealand, *The Fourth Int. Workshop on Seismic Anisotropy,* Edinburgh, (Abstract).

Goetze, C., 1969. High temperature elasticity and anelasticity of polycrystalline salts, Ph.D. thesis, Harvard University, Cambridge, Mass..

Goto, T., Ohno, I. and Sumino, Y., 1976. The determination of the elastic constants of natural almandine-pyrope garnet by rectangular parallelepiped resonance method, *J. Phys. Earth, 24,* 149-156.

Goto, T., Anderson, O., Ohno, I. and Yamamoto, S., 1989. Elastic constants of corundum up to 1825°K., *J. Geophys. Res., 94,* 7588-7602.

Grand, S. and Helmberger, D.V., 1984. Upper mantle shear structure of North America, *Geophys. J. R. astr. Soc., 76,* 399-438.

Grau, G., 1988. S-waves and seismic anisotropy, *Eötvös Lorand Geophys. Inst. of Hungary, Geophys. Trans., 34,* 7-43.

Greenhalgh, S.A., Wright, C., Goleby, B. and Soleman, S., 1990. Seismic anisotropy in granulite facies rocks of the Arunta Block, Central Australia, *Geophys. Res. Lett., 17,* 1513-1516.

Gueguen, Y. and Darot, M., 1982. Upper mantle plasticity from laboratory experiments, *Phys. Earth Planet. Int., 29,* 51-57.

Gueguen, Y. and Mercier, 1973. High attenuation and the low velocity zone, *Phys. Earth Planet. Int., 7,* 39-46.

Gueguen, Y. and Nicolas, A., 1980. Deformation of mantle rocks, *Ann. Rev. Earth Planet. Sci., 8,* 119-144.

Guggisberg, B., 1986. *Eine zweidimensionale refraktionsseismische Interpretation der Geschwindigkeits- Tiefen- Struktur des oberen Erdmantels unter dem fennoskandischen Schild (Projekt Fennolora), Diss. ETH, 7945,* Zurich, 199 pp.

Gupta, I.N., 1973a. Dilatancy and premonitory variations of P, S travel times, *Bull. Seismol. Soc. Am., 63,* 1157-1161.

Gupta, I.N., 1973b. Premonitory variations in S-wave velocity anisotropy before earthquakes in Nevada, *Science, 182,* 1129-1132.

Gutenberg, B., 1948. On the layer of relatively low wave velocity at a depth of about 80 kilometers, *Bull. Seismol. Soc. Am., 38,* 121-148.

Gutenberg, B., 1959. The asthenosphere low velocity layer, *Ann. Geofis., Rome, 12,* 439-460.

Hadiouche, O., Jobert, N. and Montagner, J.P., 1989. Anisotropy of the African continent inferred from surface waves, *Phys. Earth Planet. Inter., 58,* 61-81.

Hager, B. and O'Connell, R., 1979. Kinematic models of large-scale flow in the Earth's mantle, *J. Geophys. Res., 84,* 1031-1048.

Harkrider, D.B. and Anderson, D.L., 1962. Computation of surface wave dispersion for multilayered anisotropic media, *Bull. Seismol. Soc. Am., 52,* 321-332.

Haskell, N.A., 1953. The dispersion of surface waves on multilayered media, Bull. Seismol. Soc. Am., 43, 17-34.

Hearmon, R. F. S., 1956. The elastic constants of anisotropic materials II, *Adv. Phys., 5,* 523-382.

Hearmon, R. F. S., 1979. In: *Landolt-Bornstein, N.S., vol. III/11,* Berlin: Springer-Verlag, S1.

Helbig, K., 1958. Elastische Wellen in anisotropen Medien, Gerlands Beitr. Geophysik, 67, 177-211.

Helbig, K., 1984. Anisotropy and dispersion in periodically layered media, *Geophysics, 49*, 364-373.

Helmstaedt, H., Anderson, O.L. and Gavasci, A.T., 1972. Petrofabric studies of eclogite, spinel-websterite, and spinel-lherzolite xenoliths from kimberlite-bearing breccia pipes in southeastern Utah and northeastern Arizona, *J. Geophys. Res.,77*, 4350-4365.

Hess, H., 1964. Seismic anisotropy of the uppermost mantle under oceans, *Nature, 203*, 629-631.

Hill, L., 1952. The elastic behaviour of a crystalline aggregate, *Proc. Phys. Soc. London, A65*, 349-354.

Hirahara, K. and Ishikawa, Y., 1984. Travel time inversion for three-dimensional P-wave velocity anisotropy, *J. Phys. Earth, 32*, 197-218.

Hirn, A., 1977. Anisotropy in the continental upper mantle: possible evidence from explosion seismology, *Geophys. J. R. astr. Soc., 49*, 49-58.

Holbrook, W.S., Gajewski, D., Krammer A. and Prodehl, C., 1988. An interpretation of wide-angle compressional and shear wave data in southern Germany: Poisson's ratio and petrological implications, *J. Geophys. Res., 93*, 81-106.

Honda, S., 1986. Strong anisotropic flow in a finely layered asthenosphere, *Geophys. Res. Lett., 13*, 1454-1457.

Hood, J., Schoenberg, M. and Fryer, G.J., 1988. Modelling the oceanic crust as an anisotropic medium with orthorhombic symmetry, in *Third Int. Workshop on Seismic Anisotropy*, Berkeley, 1988, (Abstract).

Hovland, J. and Husebye, E.S., 1981. The upper mantle beneath southeastern Europe, in *Identification of Seismic Sources - Earthquake or Underground Explosion, E.S. Husebye and S. Mykkeltveit (eds.), Reidel*, 589-605.

Hudson, J.A., 1980. Overall properties of a cracked solid, Math. Proc. Camb. phil. Soc., 88, 371-384.

Hudson, J.A., 1981. Wave speeds and attenuation of elastic waves in material containing cracks, Geophys. J. R. astr. Soc., 64, 133-150.

Hudson, J.A., 1990. Overall elastic properties of isotropic materials with arbitrary distribution of circular cracks, *Geophys. J. Int., 102*, 465-469.

Humphreys, E. and Clayton, R.W., 1988. Adaptation of back projection tomography to seismic travel time problems, *J. Geophys. Res., 93*, 1073-1085.

Huntington, H. B., 1958 . The elastic constants of crystals, in *"Solid State Physics", Vol.7, F. Seitz, D. Turnbull, (eds.), Academic Press*, New York, 213-285.

Ida, Y., 1969. Thermodynamic theory of nonhydrostatically stressed solid involving finite strain, *J. Geophys. Res., 74*, 3208-3218.

Ida, Y., 1984. Preferred orientation of olivine and anisotropy of the oceanic lithosphere, *J. Phys. Earth , 32*, 245-257.

Ide, J. M., 1935. Some dynamic methods for determination of Young's modulus, *Rev. Sci. Instr, 6*, 296-298.

Isaak, D. G., Anderson, O.L. and Goto, T., 1989. Elasticity of single-crystal fosterite measured to 1700 K°, *J. Geophys. Res., 94*, 5895-5906.

Ishikawa, Y., 1984. Anisotropic plate thickening model, *J. Phys. Earth, 32*, 219-228.

Jacobsen, S.B. and Wasserburg, G.J., 1979. The mean age of mantle and crustal reservoirs, *J. Geophys. Res., 84*, 7411-7427.

Jackson, D.D., 1972. Interpretation of inaccurate, insufficient and inconsistent data, *Geophys. J. R. astr. Soc., 28*, 97-109.

Jackson, D.D., 1979. The use of a priori data to resolve non-uniqueness in linear

inversion, *Geophys. J. R. astr. Soc.*, 57, 137-157.
James, D., 1971. Anomalous Love wave phase velocities, J. Geophys. Res., 76, 2077-2083.
Jeanloz, R., 1990. The nature of the Earth's core, *Ann. Rev. Earth Planet. Sci., 18,* 357-386.
Jeanloz, R. and Thompson, A. B., 1983. Phase transitions and mantle discontinuities, *Rev. Geophys. Space Phys., 21,* 51-74.
Jeanloz, R. and Wenk, H.R., 1988. Convection and anisotropy of the inner core, *Geophys. Res. Lett., 15,* 72-75.
Jech, J. and Psencik, I., 1989. First-order perturbation method for anisotropic media, *Geophys. J. Int., 99,* 369-376.
Jephcoat, A.P., Mao, H.K. and Bell, P.M., 1986. Static compression of iron to 78 GPa with rare gas solids as pressure-transmitting media, *J. Geophys. Res., 91,* 4677-4684.
Ji, S. and Mainprice, D., 1988. Natural deformation fabrics of plagioclase: implication for slip systems and seismic anisotropy, *Tectonophys., 147,* 145-163.
Jones, T.D. and Nur, A., 1982. Seismic velocity and anisotropy in mylonites and the reflectivity of deep crustal faults, *Geology, 10,* 260-263.
Jones, T.D. and Nur, A., 1984. The nature of seismic reflections from deep crustal fault zones, *J. Geophys. Res., 89,* 3153-3171.
Journet, B. and Jobert, N., 1982. Variation with age of anisotropy under oceans from great circle surface waves, *Geophys. Res. Lett., 9,* 179-181.
Juhlin, C., 1990. *Seismic attenuation, shear wave anisotropy and some aspects of fracturing in the crystalline rock of the Siljan Ring area, central Sweden,* Ph.D. thesis, Univ. of Uppsala.
Kaarsberg, E.A., 1968. Elasticity studies of isotropic rock samples, *Trans. Soc. Min. Eng., 241,* 470-475.
Kamb, W.B., 1959. Theory of preferred crystal orientation developed by crystallization under stress, *J. Geol., 67,* 153-170.
Kanamori, H. and Anderson, D.L., 1977. Importance of physical dispersion in surface wave and free oscillation problems: a review, *Rev. Geophys. Space Phys., 15,* 105-112.
Kaneshima, S., 1990. Origin of crustal anisotropy: shear wave splitting studies in Japan, *J. Geophys. Res., 95,* 121-133.
Kaneshima, S. and Ando, M., 1988. Crustal anisotropy inferred from shear-wave splitting, *The Third Int. Workshop on Seismic Anisotropy*, Berkeley, USA, (Abstract).
Kaneshima, S., Ito, H. and Sugihara, M., 1989. Shear-wave polarization anisotropy observed in a rift zone in Japan, *Tectonophys., 157,* 281-300.
Karato, S. and Spetzler, H.H., 1990. Defect microdynamics in minerals and solid-state mechanisms of seismic wave attenuation and velocity dispersion in the mantle, *Rev. Geophys., 28,* 399-421.
Karato, S., Toriumi, M. and Fujii, T., 1980. Dynamic recrystallisation of olivine single crystals during high-temperature creep, *Geophys. Res. Lett., 7,* 649-652.
Karson, J.A. and Elthon, D., 1987. Evidence for variations in magma production along oceanic spreading centers: a critical appraisal, *Geology, 15,* 127-131.
Kawasaki, I., 1986. Azimuthally anisotropic model of the oceanic upper mantle, *Phys. Earth Planet. Inter., 43,* 1-21.
Kawasaki, I. and Konno, F., 1984. Azimuthal anisotropy of surface waves and the possible type of the seismic anisotropy due to preferred orientation of olivine in the uppermost mantle beneath the Pacific ocean, *J. Phys. Earth, 32,* 229-244.

Kearey, P. and Vine, F.J., 1990. *Global tectonics*, Blackwell Sci. Publ., Oxford.

Keen, C.E. and Barrett, D.L., 1971. A measurement of seismic anisotropy in the Northeast Pacific, *Can. J. Earth Sci., 8*, 1056-1064.

Keen, C.E. and Tramontini, C. 1970. A seismic refraction survey on the Mid-Atlantic Ridge, *Geophys. J. R. astr. Soc., 20*, 473-491.

Keilis Borok, V.I., 1989. *Seismic surface waves in a laterally inhomogeneous Earth*, Kluwer Acad. Publ., Dordrecht, 293 pp.

Keilis Borok, V.I., Levshin, A.L. and Valyus, V.P., 1969. Determination of the upper mantle; velocity cross-section for Europe, *Izv. Acad. Sci. USSR, 185, n°3*, (in Russian).

Keith, C. M. and Crampin, S., 1977. Seismic body waves in anisotropic media: synthetic seismograms, *Geophys. J.R. astr. Soc., 49*, 225-243.

Kern, H., 1974. Gefugeregelung und elastische Anisotropie eines Marmors, *Contr. Mineral. and Petrol., 43*, 47-54.

Kern, H., 1982. Elastic wave velocities and constants of elasticity of rocks at elevated pressures and temperatures, in *Landolt-Bornstein, Numerical Data and Functional Relationship in Science and Technology, Physical properties of rocks, K.H. Hellwege ed., Vol.1b*, Springer-Verlag, Berlin, Heidelberg, New York, 99-140.

Kern, H., Fakhimi, M., 1975. Effect of fabric anisotropy on compressional wave propagation in various metamorphic rocks for the range 20-700° C at 2 kbars, *Tectonophys., 28*, 227-244.

Kern, H., Schenk, V., 1985. Elastic wave velocities in rocks from a lower crustal section in nothern Calabria (Italy), *Phys. Earth Planet. Int., 40*, 147-160.

Kern, H. and Schenk, V., 1988. A model of velocity structure beneath Calabria, Southern Italy, based on laboratory data, *Earth Planet. Sci. Lett., 87*, 235-237.

Kern, H. and Wenk, H.R., 1990. Fabric related velocity anisotropy and shear-wave splitting in rocks from the Santa Rosa Mylonite Zone, California, *J. Geophys. Res., 95*, 11213-11223.

Kind, R. and Mueller, G. , 1977. The structure of the outer core from SKS amplitudes, *Bull. Seismol. Soc. Am., 67*, 1541-1554.

Kind, R., Kosarev, G.L., Makeeva, L.I. and Vinnik, L.P., 1985. Observations of laterally inhomogeneous anisotropy in the continental lithosphere, *Nature, 318*, 358-361.

Kirkwood, S.C., 1978. The significance of isotropic inversion of anisotropic surface-wave dispersion, Geophys. J. R. astr. Soc., 55, 131-142.

Kirkwood, S.C. and Crampin, S., 1981. Surface-wave propagation in an ocean basin with an anisotropic upper mantle: observations of polarization anomalies, Geophys. J. R. astr. Soc., 64, 487-497.

Knittle, E., Jeanloz, R. and Smoth, G.L., 1986. Thermal expansion of silicate perovskite and stratification of the Earth's mantle, *Nature, 319*, 214-216.

Kogan, S.D., 1981. Anisotropy and large-scale lateral inhomogeneity of the upper mantle, *Phys. Earth Planet. Int., 26*, 171-178.

Koshubin, S.I., Pavlenkova, N.I. and Yegorkin, A.V., 1984. Crustal heterogeneity and velocity anisotropy from seismic studies in the USSR, *Geophys. J.R. astr. Soc., 76*, 221-226.

Kroenke, L.W., Manghnani, M.H., Rai, C.S., Fryer, P. and Ramananantoandro, R., 1976. Elastic properties of selected ophiolite rocks from Papua New Guinea: nature and composition of oceanic lower crust and upper mantle, in *"The Geophysics of the Pacific Ocean Basin and its Margins", A.G.U. Monogr., 19*, 407-422.

Kumazawa, M., 1963. Fundamental theory on the nonhydrostatic thermodynamics and on the stability of mineral orientation and phase aquilibrium, *J. Earth Sci. Nagoya Univ., 11*, 145-217.

Kumazawa, M., 1969. The elastic constants of single-crystal orthopyroxene, *J. Geophys. Res., 74*, 5973-5980.

Kumazawa, M. and Anderson, O.L., 1969. Elastic moduli, pressure derivatives, and temperature derivatives of single-crystal olivine and single-crystal forsterite, *J. Geophys. Res., 74*, 5961-5972.

Kumazawa, M., Helmstaedt, H. and Masaki, K., 1971. Elastic properties of eclogite xenoliths from diatremes of the east Colorado plateau and their implications of the upper mantle structure, *J. Geophys. Res., 56,* 1231-1247.

Lanczos, C., 1961. *Linear differential operators*, Van Nostrand ed., 564 pp.

Landau, L. and Lifchitz, E., 1964. *Elasticity theory*, Mir Publ. House, Moscow.

Lander, A.C., 1984. Surface-wave anomalous occurrences in North East Eurasia and their relation to the Momsky rift area, in *"Mathematical Modelling and Interpretation of Geophysical Data, Computation Seismology 16,* 127-155, Nauka, Moscow, (in Russian).

Landolt-Bornstein, 1982. *Numerical Data and Functional Relationships in Science and Technology, vol. 1, Phys. Properties of Rocks*, Springer-Verlag, K.H. Hellwege (ed.), Berlin, Heidelberg, New York, 604 pp.

Langseth, M.G. and Moore, J.C., 1990. Fluids in accretionary prisms, *EOS, 71*, 245-247.

Layat, C., Clement, A.C., Pommier, G. and Buffet, A., 1961. Some technical aspects of refraction of seismic prospecting in the Sahara, *Geophysics, 26,* 437-446.

Leary, P.C., Crampin, S. and McEvilly, T.V., 1990. Seismic fracture anisotropy in the Earth's crust: An overview, *J. Geophys. Res., 95*, 11105-11114.

Lefeuvre, F., Cliet, C. and Nicoletis, L., 1989. Shear-wave birefringence measurements and detection in the Paris basin, *SEG Annual Meeting,* Dallas, 786-790.

Leveque, J.J and Cara, M., 1983. Long-period Love-wave overtone data in North America and the Pacific Ocean: New evidence for upper-mantle anisotropy, *Phys. Earth Planet. Int., 33,* 164-179.

Leveque, J.J. and Cara, M., 1985. Inversion of multimode surface-wave data: evidence for sub-lithospheric anisotropy, *Geophys. J. R. astr. Soc., 83,* 753-774.

Levien, L., Weidner, D.J. and Previtt, C.T., 1979. Elasticity of diopside, *Phys. Chem. Minerals, 4*, 105-113.

Levitova, F.M., 1980. *Anisotropy of elastic properties of rocks and the relation to seismic anisotropy of deep structure of the Earth,* PhD thesis, Institute of Physics of Earth, Moscow (in Russian).

Levshin, A. and Ratnikova, L., 1984. Apparent anisotropy in inhomogeneous media, *Geophys. J. R. astr. Soc., 76*, 65-69.

Levykin, A.I., 1974. Elastic properties of some minerals and monomineral rocks at pressures to 55 kilobars, in *Physical Properties of Rocks at High Pressures and Temperatures , Metsniereba,* Tbilisi , 90-92, (in Russian).

Li, Y.G., Leary, P.C. and Henyey, T.L., 1988. Stress orientation inferred from shear wave splitting in basement rock at Cajon Pass, *Geophys. Res. Lett., 15*, 997-1000.

Li, Y.G., Leary, P.C. and Aki, K., 1990. Ray series modeling of seismic wave travel times and amplitudes in three-dimensional heterogeneous anisotropic crystalline rock: borehole vertical seismic profiling seismograms for the Mojave Desert, California, *J. Geophys. Res., 95*, 11225-11239.

Liu, E. and Crampin, S., 1990. Effects of the internal shear wave window: comparison with anisotropy induced splitting, *J. Geophys. Res., 95,* 11275-11281.

Liu, H.L. and Helmberger, D.V., 1985. The 23:19 aftershock of the October 1979 Imperial Valley earthquake: more evidence for an asperity, *Bull. Seismol. Soc. Am., 75*, 689-708.

Lliboutry, L. and Duval, P., 1985. Various isotropic and anisotropic ices found in glaciers and ice caps and their corresponding rheologies, *Ann. Geophys., 3*, 207-224.

Love, A.E.H., 1927. *A treatise on the mathematical theory of elasticity*, Dover publ., New York, 1944, 643 pp.

Luschen, E., Nolte, B. and Fuchs, K., 1990. Shear-wave evidence for an anisotropic lower crust beneath the Black Forest, southwest Germany, *Tectonophys., 173*, 483-493.

Luschen, E., Sollner, W. and Hohrath, A., 1991. Integrated P- and S-wave borehole experiments at the KTB-deep drilling site in the Oberpfalz area (SE Germany). *Proc. of Symp. on Deep Seismic Reflection Profiling of the Continental Lithosphere, Beyreuth 1990, AGU Geodyn. Series (in press).*

Maaloe, S. and Steel, R., 1980. Mantle composition derived from the composition of lherzolites, *Nature, 285*, 321-322.

Mainprice, D. 1990. A fortran program to calculate anisotropy from the lattice preferred orientation of minerals, *Computers & Geosciences, 16*, 385-393.

Mainprice, D. and Casey, M., 1990. The calculated seismic properties of quartz mylonites with typical fabrics: relationship to kinematics and temperature, *Geophys. J. Int., 103*, 599-608.

Mainprice, D. and Nicolas, A., 1989. Development of shape and lattice preferred orientation: application to the seismic anisotropy of the lower crust, *J. Struct. Geol., 11*, 175-189.

Majer, E.L., McEvilly, T.V., Eastwood, F.S. and Myer, L.R., 1988. Fracture detection using P-wave and S-wave vertical seismic profiles at the Geysers geothermal field, *Geophysics, 53*, 76-84.

Makeeva, L.I., Plesinger, A. and Horalek, J., 1990. Azimuthal anisotropy beneath the Bohemian Massif from broad-band seismograms of SKS waves, *Phys. Earth Planet. Int., 62*, 298-306.

Malecek, S.J. and Clowes, R.M., 1978. Crustal structure near Explorer Ridge from a Marine Deep Seismic Sounding Survey, *J. Geophys. Res., 83*, 5899-5912.

Manghnani, M.H., Ramananantoandro, R. and Clark, S.P. (Jr), 1974. Compressional and shear wave velocities in granulite facies rocks and eclogites to 10 kbar, *J. Geophys. Res., 79*, 5427-5446.

Masse, R.P., 1987. Crustal and upper mantle structure of stable continental regions in North America and northern Europe, *Pageoph, 125*, 205-239.

Matsu'ura, M. and Hirata, N, 1982. Generalized least-square solutions to quasi linear inverse problems with a priori informations, *J. Phys. Earth, 30*, 451-468.

Maupin, V., 1985. Partial derivatives of surface wave phase velocities for flat anisotropic models, *Geophys. J. R. astr. Soc., 83*, 379-398.

Maupin, V., 1987. *Etude des caractéristiques des ondes de surface en milieu anisotrope: application à l'analyse d'anomalies de polarisation à la station de Port-aux-français*, Ph.D. thesis, Univ. Louis Pasteur, Strasbourg.

Maupin, V., 1989. Surface wave in weakly anisotropic structures: on the use of ordinary or quasi-degenerate perturbation methods, *Geophys. J. Int., 98*, 553-563.

Maupin, V., 1990. Modelling of three-component Lg waves in anisotropic crustal models, *Bull. Seismol. Soc. Am, 80*, 1311-1325.

Maupin, V. and Cara, M., 1991. Love- and Rayleigh-wave incompatibility and upper mantle anisotropy in the Iberian Peninsula, submitted to *Tectonophys.*.

McCarthy, J., Fuis, G. and Wilson, J. 1987. Crustal structure of the Whipple Mountains, southeastern California: a refraction study across a region of large continental extension, *Geophys. J. R. astr. Soc., 89*, 119-124.

McEvilly, T.V., 1964. Central U.S. crust-upper mantle structure from Love and Rayleigh wave phase velocity inversion, *Bull. Seismol. Soc. Am., 54*, 1997-2015.

McGeary, S. and Warner, M.R., 1985. Seismic profiling the continental lithosphere, *Nature, 317*, 795-797.

McKenzie, D., 1979. Finite deformation during fluid flow, *Geophys. J.R. astr. Soc., 58*, 689-715.

Meissner, R. and Kusznir, N.J., 1987. Crustal viscosity and the reflectivity of the lower crust, *Ann. Geophys., 5B*, 365-374.

Miller, H., 1982. Physical properties of ice, in *Landolt - Bornstein, Numerical Data and Functional Relationships in Science and Technology, vol. 1, Physical properties of ice, Springer-Verlag* Berlin, Heidelberg, New York, 482-507.

Minster, B. and Anderson, D.L., 1980. Dislocations and nonelastic processes in the mantle, *J. Geophys. Res., 85*, 6347-6352.

Mitchell, B., 1984. On the inversion of Love- and Rayleigh-wave dispersion and implication for Earth structure and anisotropy, *Geophys. J. R. astr. Soc., 76*, 233-241.

Mitchell, B.J. and Yu, G.K., 1980. Surface wave dispersion, regionalized velocity models, and anisotropy of the Pacific crust and upper mantle, *Geophys. J. R. astr. Soc., 63*, 497-514.

Mochizuki, E., 1986. The free oscillations of an anisotropic and heterogeneous earth, *Geophys. J. R. astr. Soc., 86*, 167 - 176.

Montagner, J.P., 1985. Seismic anisotropy of the Pacific Ocean inferred from long-period surface waves dispersion, *Phys. Earth Planet. Int., 38*, 28-50.

Montagner, J.P. and Anderson, D.L., 1989a. Petrological constraints on seismic anisotropy, *Phys. Earth Planet. Int., 54*, 82-105.

Montagner, J.P. and Anderson, D.L., 1989b. Constrained reference model, *Phys. Earth. Planet. Int., 58*, 205-227.

Montagner, J.P. and Jobert, N., 1988. Vectorial tomography - II. Application to the Indian Ocean, *Geophys. J., 94*, 309-344.

Montagner, J.P. and Nataf, H.C., 1986. A simple method for inverting the azimuthal anisotropy of surface waves, *J. Geophys. Res., 91*, 511-520.

Montagner, J.P. and Nataf, H.C., 1988. Vectorial tomography, -I Theory, *Geophys. J., 94*, 295-307.

Montagner, J.P. and Tanimoto, T., 1990. Global anisotropy in the upper mantle inferred from the regionalization of phase velocities, *J. Geophys. Res., 95*, 4797-4819.

Montagner, J.P. and Tanimoto, T., 1991.Global upper mantle tomography of seismic velocities and anisotropies, submitted to *J. Geophys. Res.*.

Mooney, W.D., Fuis, G.S., Ambos, E.L., Page, R.A., Plafker, G., Nokleberg, W.J. and Campbell, D.L., 1987. A mechanism for the incorporation of upper mantle material into the middle and upper crust: Evidence from southern Alaska, in *"Geophysics and Petrology of the Deep Crust and Upper Mantle", Noller, J.S. et al. (eds.), U.S. Geol. Survey Circ. 956*, 37.

Morelli, A., Dziewonski, A.M. and Woodhouse, J.H., 1986. Anisotropy of the inner core inferred from PKIKP travel times, *Geophys. Res. Lett., 13*, 1545-1548.

Morris, G.B., Raitt, R.W. and Shor, G.G., 1969. Velocity anisotropy and delay time maps of the mantle near Hawaii, *J. Geophys. Res., 74*, 4300-4316.

Mueller, S. and Ansorge, J., 1986. Long-range seismic refraction profiles in Europe, in

Reflection Seismology: a Global Perspective, M. Barazangi and L. Brown (eds.), Geodyn. Ser. AGU , 13, 167-182.

Nataf, J.C., Nakanishi, I. and Anderson, D.L., 1986. Measurements of mantle wave velocities and inversion for lateral heterogeneities and anisotropy: 3. Inversion, *J. Geophys. Res., 91,* 7261-7307.

Neunhofer, H. and Malischewski, P., 1981. Anomalous polarization of Love waves indicating anisotropy along paths in Eurasia, *Gerlands Beitr. Geophysik, 90,* 179-186.

Nicolas, A. and Christensen, N.I., 1987. Formation of anisotropy in upper mantle peridotites - A review, in *"Composition, Structure and Dynamics of the Lithosphere-Asthenosphere System", K. Fuchs and C. Froidevaux (eds.), Geodyn. Ser. AGU, 16,* 111-123.

Nicolas, A. and Poirier, J. P., 1976. *Crystalline plasticity and solid state flow in metamorphic rocks,* Wiley, London, 444 pp.

Nicolas, A., Boudier, F. and Boullier, A.M., 1973. Mechanisms of flow in naturally and experimentally deformed peridotites, *Am. Jour. Sci., 273,* 853-876.

Nishimura, C.E. and Forsyth, D.W., 1989. The anisotropic structure of the upper mantle in the Pacific, *Geophys. J. Int., 96,* 203-229.

Nolet, G., 1975. Higher Rayleigh modes in western Europe, *Geophys. Res. Lett., 2,* 60-62.

Nolet, G., 1977. The upper mantle under western Europe inferred from the dispersion of Rayleigh modes, *Z. Geophys., 43,* 265-286.

Nolet, 1990. Partitioned waveform inversion and two-dimensional structure under the network of autonomously recording seismographs, *J. Geophys. Res., 95,* 8499-8512.

Nowack, R. and Psencik, I., 1991. Perturbation from isotropic to anisotropic heterogeneous media in ray approximation, *Geophys. J. Int., 106,* in press.

Nur, A., 1971. Effects of stress on velocity anisotropy in rocks with cracks, J. Geophys. Res., 76, 2022-2034.

Nur, A., 1972. Dilatancy, pore fluids, and premonitory variations of ts/tP travel times, *Bull. Seismol. Soc. Am., 62,* 1217-1222.

Nur, A., 1988. Crustal velocity anisotropy. The opening of dilatant cracks or the closure of preexisting ones?, *The Third Int. Workshop on Seismic Anisotropy,* Berkeley, (Abstract).

Nur, A. and Simmons, G., 1969a. Stress-induced velocity anisotropy in rock: An experimental study, *J. Geophys. Res., 74,* 6667-6675.

Nur, A. and Simmons, G., 1969b. The effect of saturation on velocity in low porosity rocks, *Earth Planet. Sci. Lett., 7,* 183-193.

Nuttli, O., 1961. The effect of the Earth's surface on the S-wave particle motion, *Bull. Seismol. Soc. Am., 51,* 237-246.

Nuttli, O., 1964. The determination of S-wave polarization angles for an Earth model with crustal layering, *Bull. Seismol. Soc. Am., 54,* 1429-1440.

Nye, J.F., 1979. *Physical properties of crystals,* Claredon Press, Oxford, 322pp..

Oehm, J., Schneider, A. and Wedepohl, K.H., 1983. Upper mantle rocks from basalts of the Northern Hessian Depression (NW Germany), *Tschermaks Min. Petr. Mitt., 32,* 25-48.

Ohno, I, 1976. Free vibration of a rectangular parallelepiped crystal and its application to determination of elastic constants of orthorhombic crystals, *J. Phys. Earth, 24,* 355-379.

Okal, E., 1977. The effect of intrinsic oceanic upper-mantle heterogeneity on regionalization of long-period Rayleigh-wave phase velocities, *Geophys. J. R.*

astr. Soc., *49*, 357-370.

Oliver, J.E., 1980. Seismic exploration of the continental basement: Trends for the 1980's, in *"Continental Tectonics", Studies in Geophysics, US Nat. Acad. Sci., Washington D.C.*, 117-126.

Paterson, M.S., 1978. Experimental rock deformation - the brittle field, *Springer Verlag*, New York, 254 pp.

Peacock, S., Crampin, S., Booth, D.C. and Fletcher, J., 1988. Shear wave splitting in the Anza seismic gap, southern California: Temporal variations as possible precursors, *J. Geophys. Res.*, *93*, 3339-3356.

Peselnick, L. and Nicolas, A., 1978. Seismic anisotropy in an ophiolite peridotite: application to oceanic upper mantle, *J. Geophys. Res.*, *83*, 1227-1235.

Peselnick, L., Nicolas, A. and Stevenson, P.R., 1974. Velocity anisotropy in a mantle peridotite from Ivrea zone: Application to upper mantle anisotropy, *J. Geophys. Res.*, *79*, 1175-1182.

Phinney, R.A. and Burridge, R., 1973. Representation of the elastic-gravitational excitation of a spherical earth model by generalized spherical harmonics, *Geophys. J. R. astr. Soc.*, *34*, 451-487.

Postma, G.W., 1955. Wave propagation in a stratified medium, *Geophysics, 20,* 780-806.

Prodehl, C., 1981. Structure of the crust and upper mantle beneath central European rift system, *Tectonophys.*, *80,* 255-269.

Prodehl, C., 1984. Structure of the earth's crust and upper mantle, in *Landolt-Bornstein, New Series, v. 2, K. Fuchs and H. Stoffel (eds.), Springer,* Berlin, 97-206.

Pros, Z. and Babuska, V., 1967. A method for investigating the elastic anisotropy on spherical rocks samples, *Z. Geophys.*, *33*, 289-291.

Pros, Z. and Podrouzkova, Z., 1974. Apparatus for investigating the elastic anisotropy on spherical samples at high pressure, *Veroff. Zentral Inst. Phys. d. Erde, Postdam*, 22, 42-47.

Raitt, R.W., 1963. Seismic refraction studies of the Mendocino fracture zone, *Rep. MPL-U-23/63, Mar. Phys. Lab. Scripps Institute of Oceanography, Univ. of California, San Diego.*

Raitt, R.W., Shor, G.G. (Jr), Francis, T.J.G. and Morris, G.B., 1969. Anisotropy of the Pacific upper mantle, *J. Geophys. Res.*, *74*, 3095-3109.

Ranalli, G. and Murphy, D.C., 1987. Rheological stratification of the lithosphere, *Tectonophys.*, *132*, 281-295.

Regan, J. and Anderson, D.J., 1984. Anisotropic models of the upper mantle, *Phys. Earth Planet. Int.*, *35*, 227-263.

Reuss, A., 1929. Berechnung der Fliessgrenze von Mischkristallen auf Grund der Plastizitätsbedingung für Einkristalle, *Z. Angew. Math. Mech.*, *9*, 49-58.

Ribe, N.M., 1989. Seismic anisotropy and mantle flow, *J. Geophys. Res.*, *94,* 4213-4223.

Ringwood, A.E., 1975. *Composition and petrology of the Earth's mantle*, McGraw Hill, New York, 618 pp.

Riznichenko, Y.V., 1949. Seismic quasi-anisotropy, *Izv. Akad. Nauk SSSR, 13,* 518-544, (in Russian).

Robertson, J.D. and Carrigan, D , 1983. Radiation patterns of the shear-wave vibrator in near-surface shale, *Geophysics, 48,* 19-26.

Roden, M.F., Smith, D. and Rama Murthy, V., 1990. Chemical constraints on lithosphere composition and evolution beneath the Colorado Plateau, *J. Geophys. Res.*, *95*, 2811-2831.

Rokhlin, S.I., Bolland, T.K. and Adler, L., 1986. Reflection and refraction of elastic waves on a plane interface between two generally anisotropic media, *J. Acoust. Soc. Am., 79*, 906-918.

Romanowicz, B., 1980. A study of large-scale lateral variations of P velocity in the upper mantle beneath western Europe, *Gophys. J. R. astr. Soc., 63*, 217-232.

Ronov, A.B. and Yaroshevsky, A.A., 1969. Chemical composition of the earth's crust, in *the Earth's Crust and Upper Mantle, P.J. Hart (ed.), AGU Monograph, 13*, 37-57.

Ryall, A. and Savage, W., 1974. S wave splitting: Key to earthquake prediction?, *Bull. Seismol. Soc. Am., 64*, 1943-1951.

Ryzhova, T.V., 1964. Elastic properties of plagioclases, *Izv. Acad. Sci. USSR, Geophys. Ser., 7*, 1049-1051, (in Russian).

Ryzhova, T.V.and Aleksandrov, K.S., 1962. The elastic properties of rock-forming minerals, IV: nepheline, *Izv. Acad. Sci. USSR, Geophys. Ser., 12*, 1799-1801, (in Russian).

Ryzhova, T.V. and Aleksandrov, K.S., 1965. The elastic properties of potassium-sodium feldspars, *Izv. Acad. Sci, USSR, Earth Phys. Ser., 1*, 98-102, (in Russian).

Ryzhova, T.V., Aleksandrov, K.S. and Korobkova, V.M., 1966. The elastic properties of rock-forming minerals. Additional data on silicates, *Izv. Acad. Sci. USSR, Earth Phys. Ser., 2*, 63-65, (in Russian).

Sabatier, P.C., 1977. Positivity constraints in linear inverse problems: I) General theory, *Geophys. J. R. astr. Soc., 48*, 443-459.

Sachpazi, M. and Hirn, A., 1991.Shear-wave anisotropy across the geothermal field of Milos, Aegean volcanic arc, submitted to *Geophys. J. Int.*.

Savage, M.K., Silver, P.G. and Meyer, R.P., 1990. Observations of teleseismic shear-wave splitting in the Basin and Range from portable and permanent stations, *Geophys. Res. Lett., 17*, 21-24.

Sawamoto, H., Weidner, D.J., Sasaki, S. and Kumazawa, M., 1984. Single-crystal elastic properties of the modified spinel (beta) phase of Mg_2SiO_4, *Science, 224*, 749-751.

Schlue, J.W. and Knopoff, L., 1976. Shear-wave anisotropy in the mantle of the Pacific Basin, *Geophys. Res. Lett., 3*, 359-362.

Schlue, J.W. and Knopoff, L., 1977. Shear wave polarization anisotropy in the Pacific Basin, *Geophys. J. R. astr. Soc., 49*, 145-165.

Schmid, S.M., Panozzo, R. and Bauer, S., 1987. Simple shear experiments on calcite rocks: rheology and microfabric, *J. Struct. Geol., 9*, 747-778.

Scholz, C., Sykes, L. and Aggarwal, Y., 1973. Earthquake prediction: A physical basis, *Science, 181*, 803-810.

Schulze, D.J., 1989. Constraints on the abundance of eclogite in the upper mantle, *J. Geophys. Res., 94*, 4205-4212.

Schwerdtner, W.M., 1964. Preferred orientation of hornblende in a banded hornblende gneiss, *Am. J. Sci., 262*, 1212-1219.

Shearer, P.M. and Orcutt, J.A., 1985. Anisotropy in the oceanic lithosphere - theory and observations from the Ngendi seismic refraction experiment in the south-west Pacific, *Geophys. J. R. astr. Soc., 80*, 493-526.

Shearer, P. M. and Toy, K. M., 1991. PKP(BC) versus PKP(DF) Differential Travel Times and Aspherical Structure in the Earth's Inner Core, *J. Geophys. Res., 96*, 2233-2247.

Shearer, P.M., Toy, K.M. and Orcutt, J.A., 1988. Axi-symmetric Earth models and inner-core anisotropy, *Nature, 333*, 228-232.

Shimamura, H., 1984. Anisotropy of the oceanic litosphere of the northwestern Pacific Basin, *Geophys. J.R. astr. Soc., 76*, 253-260.

Shimamura, H., Asada, T., Suyehiro, K., Yamada, T. and Inatani, H., 1983. Longshot experiments to study velocity anisotropy in the oceanic lithosphere of the northwestern Pacific, *Phys. Earth Planet. Inter., 31*, 348-362.

Shor, G.G. (Jr.) and Pollard, D.D., 1964. Mohole site selection studies north of Mani, *J. Geophys. Res., 69*, 1627-1627.

Shor, G.G. (Jr.), Raitt, R.W., Henry, M., Bentley, L.R. and Sutton, G.H., 1973. Anisotropy and crustal structure of the Cocos Plate, *Geofis. Int., 13*, 337-362.

Siegesmund, S., Takeshita, T. and Kern, H., 1989. Anisotropy of VP and VS in an amphibolite of the deeper crust and its relationship to the mineralogical, microstructural and textural characteristics of the rock, *Tectonophys., 157*, 25-38.

Silver, P.G. and Chan, W.W., 1988. Implications for continental structure and evolution from seismic anisotropy, *Nature, 335*, 34-39.

Silver, P.G. and Chan, W.W., 1991. Shear wave splitting and subcontinental mantle deformation, *J. Geophys. Res.*, in press.

Simmons, G., 1964. Velocity of shear waves in rocks to 10 kilobars, 1, *J. Geophys. Res., 69*, 1123-1130.

Simmons, G. and Wang, H., 1971. *Single Crystal elastic constants and calculated aggregate properties: A handbook*, M.I.T. Press, Cambridge.

Sleep, N.H. and Rosendahl, B.R., 1979. Topography and tectonics of mid-oceanic ridge axes, *J. Geophys. Res., 84*, 6831-6839.

Smith, M.L. and Dahlen, F.A., 1973. The azimuthal dependence of Love and Rayleigh wave propagation in a slightly anisotropic medium, *J. Geophys. Res., 78*, 3321-3333.

Smithson, S.B., 1989. Contrasting types of lower crust, in *"Properties and Processes of Earth's Lower Crust", Mereu, R.F. et al. (eds.), AGU Geophys. Monograph, 51, Washington D.C.*, 53-63.

Smithson, S.B., Johnson, R.A., Hurich, C.A., Valasek, P.A. and Branch, C., 1987. Deep crustal structure and genesis from contrasting reflection patterns: an integrated approach, *Geophys. J. R. astr. Soc., 89*, 67-72.

Snydsman, W.E., Lewis, B.T.R. and McClain, J.M., 1975. Upper mantle velocities on the northern Cocos plate, *Earth Planet. Sci. Lett., 28*, 46-50.

Souriau, A. and Souriau, M., 1989. Ellipticity and density at the inner core boundary from subcritical PKiKP and PcP data, *Geophys. J. Int., 98, 39-54*.

Spakman, W., Wortel, M.J.R. and Vlaar, N.J., 1988. The Hellenic subduction zone: a tomographic image and its geodynamic implications, *Geoph. Res. Lett., 15*, 60-63.

Stephen, R.A., 1981. Seismic anisotropy observed in upper oceanic crust, *Geophys. Res. Lett., 8*, 865-868.

Stephen, R.A., 1985. Seismic anisotropy in the upper oceanic crust, *J. Geophys. Res., 90*, 11383-11396.

Stoneley, R., 1949. The seismological implications of aeolotropy in continental structures, *Month. Not. R. astr. Soc., Geophys. suppl. 5*, 222-232.

Stoneley, R., 1955. The propagation of surface elastic waves in a cubic crystal, *Proc. R. Soc. A., 232*, 447-458.

Stoneley, R., 1963. The propagation of surface waves in an elastic medium with orthorhombic symmetry, *Geophys. J. R. astr. Soc., 8*, 176-186.

Stuzmann, E., 1990. Etude de l'anisotropie azimutale sur sismogrammes synthétiques et données réelles en sismique de puits, *EOPGS Engineering Diploma,* University of Strasbourg.

Suda, N. and Fukao, Y., 1990. Structure of the inner core inferred from observations of seismic core modes, *Geophys. J. Int., 103*, 403-413.

Sugimura, A. and Uyeda, S., 1967. A possible anisotropy of the upper mantle accounting for deep earthquake faulting, *Tectonophys., 5*, 25-33.

Synge, J.L., 1957. Elastic waves in anisotropic media, J. Math. Phys., 35, 323-334.

Takeuchi, H. and Saito, M., 1972. *Seismic surface waves. In Mehods in Computational Physics, B.A. Bolt (ed.), Acad. Press, New York, 11*, 217-295.

Takeuchi, H., Press, F. and Kobayashi, N., 1959. Rayleigh-wave evidence for the low-velocity zone in the mantle, *Bull. Seismol. Soc. Am., 49*, 355-364.

Tanimoto, T. and Anderson, D.L., 1984. Mapping convection in the mantle, *Geophys. Res. Lett., 11*, 287-290.

Tanimoto, T. and Anderson, D.L., 1985. Lateral heterogeneity and azimuthal anisotropy of the upper mantle: Love and Rayleigh waves 100-250 s, *J. Geophys. Res., 90*, 1842-1858.

Tarantola, 1987. *Inverse problem theory, methods for data fitting and model parameter estimation*, Elsevier, 613 pp.

Tarantola, A. and Valette, B., 1982. Generalized nonlinear inverse problems solved using the least squares criterion, *Rev. Geophys. Space Phys., 20*, 219-232.

Thill, R.E., Willard, R.J. and Bur, T.R., 1969. Correlation of longitudinal velocity variation with rock fabric, *J. Geophys. Res., 74*, 4897-4909.

Thomsen, L., 1986. Weak elastic anisotropy, *Geophysics, 51*, 1954-1966.

Thomson, W.T., 1950. Transmission of elastic waves through a stratified solid medium, *J. Appl. Phys., 21*, 89-93.

Tilmann, S.E. and Bennet, H.F., 1973. Ultrasonic shear wave birefringence as a test of homogenous elastic anisotropy, *J. Geophys. Res., 78*, 2623-2629.

Todd, T., Simmons, G. and Baldridge, W.S., 1973. Acoustic double refraction in low-porosity rocks, *Bull. Seismol. Soc. Am., 63*, 2007-2020.

Toksoz, N. and Anderson, D.L., 1963. Generalized two-dimensional model seismology with application to anisotropic Earth models, *J. Geophys. Res., 68*, 1121-1130.

Trampert, J. and Leveque, J.J., 1990. SIRT: physical interpretation based on the generalized least-square solution, *J. Geophys. Res.*, 95, 12553-12559.

Tullis, T., 1971. *Experimental development of preferred orientations of mica during recrystallization*, Ph.D. Thesis, Univ. of Calif., Los Angeles.

Tullis, J., 1979. High temperature deformation of rocks and minerals, *Rev. Geophys. Space Phys., 17*, 1137-1154.

Uhrig, L.F. and Van Melle, F.A., 1955. Velocity anisotropy in laminated media, *J. Acoust. Soc. Am., 27*, 310-317.

Van der Sluis, A. and Van der Vorst, H.A., 1987. Numerical solution of large, sparse linear algebraic systems arising from tomographic problems, in *Seismic tomography, G. Nolet (ed.), Reidel Co.*, 49-83.

Vavrycuk, V., 1991. Polarization properties of elastic waves in isotropic and anisotropic media, *Ph.D. thesis, Geophysical Institute, Czechosl. Acad. Sci., Prague* (in Czech).

Vauchez, A. and Nicolas, A., 1991. Mountain building: strike-parallel motion and mantle anisotropy, *Tectonophys.*, 185, 183-201.

Verma, R.K., 1960. Elasticity of some high-density crystals, *J. Geophys.Res., 65*, 757-766.

Vetter, U. and Minster, J.B., 1981. Pn velocity anisotropy in southern California, *Bull. Seismol. Soc. Am., 71*, 1511-1530.

Vinnik, L.P. and Yegorkin, A.V., 1980. Wave fields and lithosphere-asthenosphere models according to data from seismic observations in Siberia, *Dokl. Akad.*

Nauk. SSSR, 250, 318-323.

Vinnik, L.P., Kosarev, G.L. and Makeyeva, L.I., 1984. Anisotropy of the lithosphere from the observation of SKS and SKKS, *Proc. Acad. Sci. USSR, 278,* 1335-1339, (in Russian).

Vinnik, L.P., Farra, V. and Romanowicz, B., 1989a. Azimuthal anisotropy in the Earth from observations of SKS at Geoscope and NARS bradband stations, *Bull. Seismol. Soc. Am., 79,* 1542-1558.

Vinnik, L.P., Farra, V. and Romanowicz, B., 1989b. Observational evidence for diffracted S_V in the shadow of the Earth's core, *Geophys. Res. Lett., 16,* 519-522.

Vinnik, L.P., Kind, R., Kosarev, G.L. and Makeyeva, 1989c. Azimuthal anisotropy in the lithosphere from observation of long period S waves, *Geophys. J. Int., 99,* 549-559.

Vlaar, N.J.,1968. Ray theory for an anisotropic inhomogeneous medium, *Bull. Seismol. Soc. Am., 58,* 2053-2072.

Voigt, W., 1928. *Lehrbuch der Kristallphysik,* B.G. Tubner, Leipzig.

Volarovich, M.P., Bayuk, E.I. and Efimova, G.A., 1975. *Elastic properties of minerals at high pressures,* Nauka, Moscow, (in Russian).

Volarovich, M.P., Bayuk, E.I. and Shaginian, G.Sh., 1977. Investigation of velocity and attenuation of longitudinal waves in rocks and minerals at pressures to 15 kbar and temperatures to 600°C, *Izv. Akad. Nauk USSR, Fiz. Zemli, 7,* 82-89, (in Russian).

Waff, H.S., 1980. Effects of the gravitational field on liquid distribution in partial melts within the upper mantle, *J. Geophys. Res., 85,* 1815-1825.

Wang, Y., Guyot, F., Yeganeh-Haeri, A. and Lieberman, R.C., 1990. Twinning in $MgSiO_3$ perovskite, *Science, 248,* 468-471.

Warner, M. and McGeary, S., 1987. Seismic reflection coefficients from mantle fault zones, *Geophys. J. R. astr. Soc., 89,* 223-230.

Watts, A.B., Bodine, J.H. and Steckler, S., 1980. Observation of flexure and the state of stress in the oceanic lithosphere, *J. Geophys. Res., 85,* 6369-6376.

Wedepohl, K.H., 1987. Kontinentaler Intraplatten-Vulkanismus am Beispiel der tertiaren Basalte der Hessischen Senke, *Fortschr. Miner., 65,* 19-47.

Weidner, D.J., 1975. Elasticity of microcrystals, *Geophys. Res. Lett., 2,* 189-192.

Weidner, D.J., 1985. A mineral physics test of a pyrolite mantle, *Geophys. Res. Lett., 12,* 417-420.

Weidner, D.J., Bass, J.D., Ringwood, A.E. and Sinclair, W., 1982. The single-crystal elastic moduli of stishovite, *J. Geophys. Res., 87,* 4740-4746.

Weidner, D.J., Savamoto, H., Sasaki, S. and Kumazawa, M., 1984. Single-crystal elastic properties of the spinel phase of Mg_2SiO_4, *J. Geophys. Res., 89,* 7852-7860.

Wenk, H.R., 1985. Measurement of pole figures, in *Preferred orientation in deformed metals and rocks: an introduction to modern texture analysis,Wenk, H.R. (ed.), Acad. Press,* Orlando, 610 pp.

Wenzel, F., Sandmeier, K.J. and Waelde, W., 1987. Properties of the lower crust from modelling refraction and reflection data, *J. Geophys. Res., 92,* 575-583.

White, R.S. and Whitmarsh, R.B., 1984. An investigation of seismic anisotropy due to cracks in the upper oceanic crust at 45°N, Mid-Atlantic Ridge, *Geophys. J. R. astr. Soc., 79,* 439-468.

Wielandt, E., Plesinger, A., Sigg, A. and Horalek, J., 1987. Deep structure of the Bohemian Massif from phase velocities of Rayleigh and Love waves, *Stud. Geoph. Geod., 31,* 1-7.

Wiggins, R.A., 1972. The general inverse problem: implication of surface waves and free oscillation for earth structure., *Rev. Geophys. Space Phys., 20*, 219-232.

Woodhouse, J.H. and Dziewonski, A.M., 1984. Mapping the upper mantle: Three dimensional modelling of Earth structure by inversion of seismic waveforms, *J. Geophys. Res., 89*, 5953-5986.

Woodhouse, J.H., Giardini, D. and Xiang-Dong L., 1986. Evidence for inner core anisotropy from free oscillations, *Geophys. Res. Lett., 13*, 1549-1552.

Yanovskaya, T.B., Panza, G.F., Ditmar, P.D., Suhadolc, P. and Mueller, S., 1990. Structural heterogeneity and anisotropy based on 2-D phase velocity patterns of Rayleigh waves in Western Europe, *Atti dell Accademia Nazionale dei Lincei, Rendiconti, Classe di Scienze fisiche, matematiche e naturali, 9*, 127-135.

Yeganeh-Haeri, A., Weidner, D.J. and Ito, E., 1989. Elasticity of $MgSiO_3$ in the perovskite structure, *Science, 243*, 787-789.

Young, C. and Lay, T., 1990. Multiple phase analysis of the shear-wave velocity structure in the D" region beneath Alaska, *J. Geophys. Res., 95*, 17385-17402.

Yu, G.K. and Mitchell, B.J., 1979. Regionalized shear velocity models of the Pacific upper mantle from observed Love and Rayleigh wave dispersion, *Geophys. J. R. astr. Soc., 57*, 311-341.

Zoback, M.D., et al., 1987. New evidence on the state of stress of the San Andreas fault system, *Science, 238*, 1105-1111.

Zollo, A. and Bernard, P., 1989. S-wave polarization inversion of the 15 October 1979, 23:19 Imperial Valley aftershock: evidence for anisotropy and a simple source mechanism, *Geophys. Res. Lett. 16*, 1047-1050.

INDEX

MODERN APPROACHES IN GEOPHYSICS

formerly *Seismology and Exploration Geophysics*

1. E.I. Galperin: *Vertical Seismic Profiling and Its Exploration Potential.* 1985 ISBN 90-277-1450-9
2. E.I. Galperin, I.L. Nersesov and R.M. Galperina: *Borehole Seismology.* 1986 ISBN 90-277-1967-5
3. Jean-Pierre Cordier: *Velocities in Reflection Seismology.* 1985 ISBN 90-277-2024-X
4. Gregg Parkes and Les Hatton: *The Marine Seismic Source.* 1986 ISBN 90-277-2228-5
5. Guust Nolet (ed.): *Seismic Tomography.* 1987 ISBN 90-277-2521-7
6. N.J. Vlaar, G. Nolet, M.J.R. Wortel and S.A.P.L. Cloetingh (eds.): *Mathematical Geophysics.* 1988 ISBN 90-277-2620-5
7. J. Bonnin, M. Cara, A. Cisternas and R. Fantechi (eds.): *Seismic Hazard in Mediterranean Regions.* 1988 ISBN 90-277-2779-1
8. Paul L. Stoffa (ed.): *Tau-p: A Plane Wave Approach to the Analysis of Seismic Data.* 1989 ISBN 0-7923-0038-6
9. V.I. Keilis-Borok (ed.): *Seismic Surface Waves in a Laterally Inhomogeneous Earth.* 1989 ISBN 0-7923-0044-0
10. V. Babuska and M. Cara: *Seismic Anisotropy in the Earth.* 1991 ISBN 0-7923-1321-6

KLUWER ACADEMIC PUBLISHERS – DORDRECHT / BOSTON / LONDON

9 780792 313212